KENT COUNTY LIBRARY
ENRICHMENT COLLECTION

WITHDRAWN

P9-CRE-100

Man and the Stars

MAN
AND THE
STARS

Hanbury Brown, FRS

1478
1978

Oxford University Press

Oxford University Press, Walton Street, Oxford OX2 6DP

OXFORD LONDON GLASGOW

NEW YORK TORONTO MELBOURNE WELLINGTON

KUALA LUMPUR SINGAPORE JAKARTA HONG KONG TOKYO

DELHI BOMBAY CALCUTTA MADRAS KARACHI

IBADAN NAIROBI DAR ES SALAAM CAPE TOWN

British Library Cataloguing in Publication Data

Brown, Robert Hanbury
   Man and the stars.
   1. Astronomy—Social aspects
   I. Title
   301.24'3   QB47   78-40237

   ISBN 0-19-851001-2

Frontispiece: The celestial globe is English and late
18th or early 19th century; the paper gores were
printed from engraved plates and coloured by
hand. From the Museum of the History of Science,
University of Oxford, photographed by M. Dudley,
AIIP, FRPS. The background shows the Crab nebula,
photograph courtesy Hale Observatories.

*Printed in Great Britain by BAS Printers Limited,
Over Wallop, Hampshire*

Preface

For fifteen years I have run an astronomical observatory in the Australian bush—Narrabri Observatory in New South Wales. In that time I must have shown the place to thousands of visitors and tried to explain what we were doing and why. This was not always easy; quite apart from the fact that most of our visitors had just travelled a long way on dusty roads in great heat, we were not doing the sort of thing which they expected to find in an observatory. We were not looking at the moon or the planets and there were no beautiful pictures of the sky. We were in fact doing something unusual—we were measuring the angular diameters of bright stars by a novel technique, intensity interferometry.

All one could hope to do during a short visit was to talk to visitors for about twenty minutes about what we were doing and point out how the work contributed to our general understanding of stars. There was no hope of explaining how an intensity interferometer works; that is hard enough with an audience of astronomers! After a short talk it was then best to let people wander about, take photographs, and ask questions.

As a subject for photographs the observatory was an unqualified success—the equipment was strikingly photogenic—and over the years we developed satisfactory answers to nearly all the questions which were asked. However, there was one particular question which always worried me because I couldn't answer it satisfactorily in the time available, usually only a few minutes before the coach left. It was often asked tentatively, away from the main group of visitors, as though it might be thought silly or, perhaps, impolite. It was always the same basic question—'This work may be competent and interesting, but is it any use?'

I found it comparatively easy, given a few minutes, to persuade people that our work was relevant to the general study of astronomy, but I have never been able to abbreviate, to my satisfaction, the arguments that astronomy is an integral part of science and that science is an essential part of our civilization. And so, as the coaches vanished in a cloud of red dust, I was often left wondering what the visitors really thought of our observatory. Many of them, I suspect, felt much the same as they would have done after a tour of a monastery—interesting maybe, but what a useless way to spend one's life!

Whether that was true or not I don't know, but I would certainly have liked to have had more time, not to explain our work in great detail, but to talk about the relevance of astronomy in the broadest sense. But there never was enough time and so, instead, I have tried to write in this book much of what I would have liked to have said. I hope it will help to answer some of the visitors whom I failed to satisfy and anyone else who is interested in the same question.

I thank the Oxford University Press for prompting the book, and I am particularly indebted to Dr John Roche and Professor Zdeněk Kopal for reading it before publication.

Sydney 1977   HANBURY BROWN

Contents

List of illustrations

Aztec calendar stone. The Aztecs studied astronomy, fixing the period of their festivals by the movements of the planets. This relief carving is about 13 feet (4 m) in diameter and weighs 20 tonnes. In the centre is the face of the sun-god, and the symbols in the band surrounding this represent previous eras. The 20-day names of the Aztec calendar are in the next band. Two great snakes, symbolizing time, form the outermost border.

1 Stars and the Calendar

Familiarity breeds contempt—or, more often, indifference. Our calendar is one of those things we take for granted like the alphabet or the use of the number zero. If they weren't there, then surely someone would soon invent them.

At first sight, making a calendar looks simple. All you have to do is to find a pretty picture and put underneath it neat rows of numbers against the days of the week, which run in an unbroken sequence from year to year. Of course, you must remember that some months have 30 days and some 31, but surely everyone knows that. Admittedly leap year is a minor complication. You must remember that there are 29 days in February in years which are divisible by 4 and that there is no extra day when leap year is a centurial year (i.e. divisible by 100) but is not divisible by 400. These rules about leap year are a bit mysterious and you may wonder if they are really necessary; perhaps they are just quaint old customs which have not yet been swept away by progress.

I first realized that the whole subject of calendars had unsuspected depths when, surreptitiously, I read the *Book of Common Prayer* during a long sermon in church. To my surprise I found that the date of Easter is not chosen by some Government agency to suit the convenience of holiday-makers or the Budget. It is, in fact, given for all time by a formula which reads like a magic spell. To cast this spell you must invoke the golden number, the dominical letter, and finally the day of the paschal full moon! What do these beautiful and mysterious words mean? The answer is surprisingly complicated and leads us to the long history of man's efforts to reconcile the rhythms of the solar system with his need to record the passage of time.

What are these rhythms? The simplest and most basic is the day; in fact it is so fundamental that we carry it with us wherever we go. It is built into our bodies as the circadian rhythms of body temperature, blood pressure, and so on. It is therefore rather surprising that, unlike the Greeks, we have no unambiguous word for a complete cycle of light and dark, and we must use the word day to mean day plus night. To be more precise we shall use the word *day* to mean *solar day*, meaning one complete rotation of Earth with respect to the sun. We must take care about this because there is another sort of day, the *sidereal day*, which marks one complete rotation of

Earth with respect to the stars (see Chapter 2). We must also decide when one day starts and the previous one ends. This is only a matter of convention and there have been calendars which started the day at sunrise, like the Greeks', or at sunset, like the Babylonians', or at noon, like astronomers' before they changed their day to start at midnight in A.D. 1925. We, like the ancient Egyptians, start our day at midnight.

The second major rhythm of the solar system is the *lunar month*. The moon is such a conspicuous timekeeper that most primitive communities based their calendars on it, and the phases of the moon inevitably governed most religious observances. To this day the major religious calendars, Moslem, Hebrew, and Christian are based on the moon. There have also been good practical reasons for keeping the moon in the calendar; it gives light to the hunter and the traveller and it governs the tides.

Thirdly, there is the rhythm of the year which is marked by the seasons and which can be followed at night by the changing pattern of the stars and, in the daytime, by changes in the path of the sun. Since earliest times primitive communities have used the direction of the rising or setting sun, or the positions of certain bright stars at sunrise or sunset, to mark important seasonal events, the time for sowing or gathering a crop, or the time to hunt or fish. As a matter of interest, if you look for examples of control of the calendar by the use of the sun you will find them far from the equator, like Stonehenge, where seasonal changes in the path of the sun are very noticeable. On the other hand, you will find the use of bright stars in countries near the equator where seasonal changes in the path of the sun are less obvious and clear skies make it easier to watch the rising and setting of stars. One of the best known examples was in ancient Egypt where the first rising of Sirius in the morning sky was used to control the start of the year at the season of inundation by the Nile.

Although it is not displayed by anything so dramatic as the changing phases of the moon, the rhythm of the year is far more significant in human affairs than is the lunar month. Thus, as organized societies have developed, they have sought to keep their calendars in step with the seasons and hence with the agricultural year. Calendars based on the moon were supported by an earlier religious tradition and, as secular interests grew stronger, most of them were displaced by calendars based on the year.

Finally, there is the week which I have left to the last because it is the only feature of the calendar which is not obviously based on some rhythm of the solar system. There have been 'weeks' of various length in history, there was an eight-day week in Rome and in comparatively recent times the decimal enthusiasts of the French revolution managed to introduce a ten-day week into France, but it lasted no more than a few years. The seven-day week is firmly entrenched and no one knows for certain where it came from. According to the Book of Exodus, Moses was given clear instructions on Mount Sinai about the length of the week which commemorates the six days it took to make Heaven and Earth, plus one rest day. One thing is certain, that the days of our modern week were named from the old astrological idea that each hour of the day is ruled by one of the heavenly bodies taken in sequence, each day being named after the body which rules its first hour. The names of the days of our week were derived by combining an order of the heavenly bodies used in ancient Greece with a 24-hour day which came originally from ancient Egypt; together they led to the sequence Sun, Moon, Mars, Mercury, Jupiter, Venus, Saturn. This original sequence can easily be seen in the Latin names for the days—Dies: Solis, Lunae, Martis, Mercurii, Jovis, Veneris, and Saturni. In English their origin is not so obvious because we have retained the old association with the heavenly bodies only for Sunday, Monday, and Saturday, and we have changed the names of the other days to those of northern gods—Tiw, Woden, Thor, and Frigg.

One cannot help wondering, however, whether the seven-day week was originally related to the moon. Seven days is the nearest we can get to dividing the lunar month into four equal parts without using fractions of a day. To keep such a week in step with the moon would have been awkward because one or two odd days would have been left over every month; nevertheless, there is some evidence that this was actually done in some earlier civilizations. Perhaps the week was originally used to mark the phases of the moon and we have forgotten the connection.

These, then, are the basic rhythms of the solar systems; how best can they be fitted into a practical system of recording time, or, in other words, a calendar? Basically the function of a calendar is to count intervals of time in such a way that every moment, past, present, or future can be identified unambiguously in a way which is generally agreed upon and understood. To do this we must chop time up into units which are numbered in some convenient way. The basic unit must be an interval which everyone can recognize and, for obvious reasons, all societies have chosen the solar day. But when we come to choose the longer units of time, things are not so straightforward. Ideally a calendar would be based on the three major rhythms of the solar system, the solar day, the lunar month, and the year, but centuries of observation and experiment have shown that these three units cannot be fitted into one simple, perfect calendar. A compromise has to be made which depends upon what we would now call the social priorities, the competing demands of religion, agriculture, government, and so on. Making this compromise is a difficult business and depends on obtaining remarkably accurate astronomical data. For this reason, governments have always consulted astronomers (and astrologers) about how to make a calendar and have financed observatories in order to gather the necessary data. For centuries the practice of astronomy was supported largely by its practical application to the calendar and, bearing in mind the difficulty of extracting money from any government for scientific research,

3

the problems must have been far more difficult than they look at first sight. What, in fact were they?

The astronomical problems of the calendar

The fundamental problem is that the major 'natural' units of the calendar, the day, month, and year are not simple, integral, fractions of each other. Consider first the relation between the day and the month. The *solar day* is the time it takes for the Earth to rotate once relative to the sun, and the month is governed by the time it takes for the moon to orbit the Earth. If we define the month as the interval between identical phases of the moon, technically called the *synodic month*, we find that it varies from 29 to 30 solar days and has an average length, not exactly $29\frac{1}{2}$ days, but 29·5306 days. Clearly we are going to have difficulty in fitting the day into the month.

Things are just as tricky when we try to fit the day into the year. First of all we must be quite clear what we mean by a year. As everyone knows, the length of the year is governed by the time it takes for the Earth to make one complete trip round the sun. But when we look into this in detail, it is not as simple as it looks. The axis of the Earth is not at right angles to the plane of its orbit but is inclined at $23\frac{1}{2}°$ and because of this the sun appears to follow an annual path in the sky, the ecliptic, which is inclined to the celestial

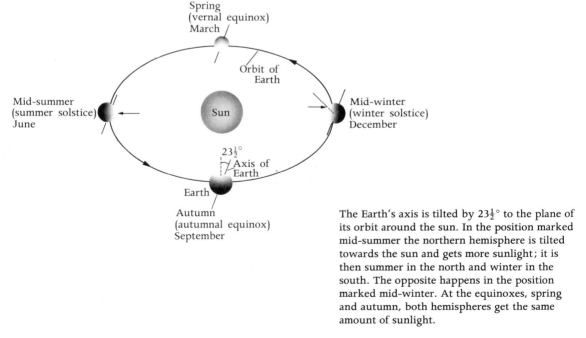

The Earth's axis is tilted by $23\frac{1}{2}°$ to the plane of its orbit around the sun. In the position marked mid-summer the northern hemisphere is tilted towards the sun and gets more sunlight; it is then summer in the north and winter in the south. The opposite happens in the position marked mid-winter. At the equinoxes, spring and autumn, both hemispheres get the same amount of sunlight.

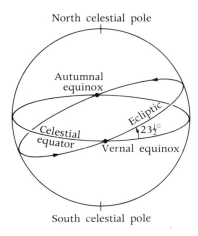

North celestial pole

Autumnal equinox

Ecliptic

$23\frac{1}{2}°$

Celestial equator

Vernal equinox

South celestial pole

The ecliptic, or the apparent path of the sun in the sky, is inclined at an angle of about $23\frac{1}{2}°$ to the celestial equator. The ecliptic and the equator intersect at the vernal and autumnal equinoxes. The direction of the sun's annual motion is shown by the arrows on the ecliptic.

equator by $23\frac{1}{2}°$. The two points where the ecliptic crosses the celestial equator are called the equinoctial points because when the sun is in those positions, day and night are of equal length. The position of an object in the sky, such as a star, is usually measured with respect to the celestial equator. Its angular distance along the equator is called *right ascension* (see Chapter 3) and is measured from the equinoctial points where the sun's path along the ecliptic crosses the celestial equator in the spring (the vernal equinox).

We are told that the Greek astronomer Hipparchos (190–120 B.C.) noticed that the positions of bright stars, as measured in his day, differed from those measured about 170 years before. Hipparchos concluded that the equinoctial points are not fixed from year to year but move at least 1° per century in a direction opposite to that of the sun's path along the ecliptic. If the equinoctial points move, then the measured positions of all the stars must also move. Hipparchos was right; the equinoctial points do, in fact, move. But two things had to be understood before this could be explained—the law of gravity and the shape of the Earth—and neither was discovered until the 17th century. It was Newton who showed that the basic reason for this movement is that the Earth is not perfectly spherical but bulges very slightly at the equator. The gravitational forces of the sun and moon, pulling on this bulge, try to twist the axis of the Earth upright, i.e. at right angles to its orbital plane. Now if you do this to a spinning top the axis remains tilted but starts to *precess* round in a circle. The Earth behaves precisely like a spinning top; its axis remains tilted at $23\frac{1}{2}°$, but precesses round in a circle. On the celestial sphere this circle has a radius of $23\frac{1}{2}°$ and the period of one complete revolution is 26 000 years. As a consequence all the stars appear to move slowly on the celestial sphere by an amount which depends in a simple geometrical way on their position in the sky. For the same reason the equinoctial points move slowly along the celestial equator by about 50 seconds of arc per year.

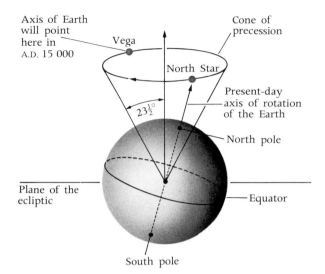

Axis of Earth will point here in A.D. 15 000

Vega

Cone of precession

North Star

Present-day axis of rotation of the Earth

$23\frac{1}{2}°$

North pole

Plane of the ecliptic

Equator

South pole

The Earth bulges very slightly at the equator. The gravitational pull of the sun and the moon try to twist the axis of the Earth so that it is at right angles to the plane of the ecliptic. This makes the axis of the Earth precess around in a circle. It takes about 26 000 years to complete one revolution. At the present time the north pole points at the Pole star; in A.D. 15000 it will point at the bright star Vega.

This phenomenon is called the *precession of the equinoxes* and must be taken into account in designing a practical calendar. Thus the year can be defined in two different ways, either as the time it takes for the sun to return to the same position in the sky relative to the stars (sidereal year) or as the time it takes for the sun to return to the vernal equinox (tropical year). Because of the precession of the vernal equinox these two years differ by approximately 20 minutes and so, in 72 years, they differ by one complete day. For a successful civil calendar the year must be related to the seasons and hence we must choose the tropical year and not the sidereal year.

Thus our basic problem is to design a calendar which keeps the calendar year as closely as possible in step with the tropical year. But again we meet the difficulty that the tropical year and solar day are awkwardly related; there are not, as many people think, exactly $365\frac{1}{4}$ days in the year but 365·2422 days and so the day will not fit precisely into the year.

Finally, there is the problem of fitting the month into the year and that, of course, is just as awkward. By dividing the length of the month into the year we find there are 12·3683 (synodic) months in one (tropical) year.

And so we see the basic astronomical problem is that the day won't fit exactly into the month and neither the day nor the month will fit into the year.

Now none of these awkward fractions would matter much if we were only concerned to make a calendar which would last for a year or two. For example, if we decided to make one year equal to 12 moons, so that our months would keep in step exactly with the moon, then our year would be 11 days shorter than a true solar year. And so the date of mid-summer for example, would slip back in the calendar by 11 days each year. This might not matter much for two or three years, but in about 16 years mid-summer and mid-winter would have exchanged dates. The trouble is that quite small errors in the design of a calendar accumulate over several years to form serious errors and, if we want it to last for centuries, then we must pay attention to these minor errors.

The history of the calendar is therefore one of the successive attempts to reconcile these awkward relations between the day, month, and year with the needs of society, and of trying to establish these relations with greater and greater precision. An enormous effort has been expended in devising calendars and in recording the movements of the sun, moon, planets, and some of the bright stars. The observational work was carried on in primitive observatories, mostly using large stone monuments and without the benefit of the telescope or clock, for thousands of years. For example, Jai Singh's observatory at Jaipur, built in the 18th century, has much in common with Stonehenge, built probably about 4000 years earlier. No doubt the principal reason for building these massive structures was to collect data for astrology and divination, but we must remember that those arts were closely connected with both government and control of the calendar and thus, in its early days, the purpose and development of astronomy was linked closely to the calendar. Our own calendar is the end-product of centuries of observation and experiment and is one of the great achievements of astronomy before the invention of the telescope. To illustrate this, let us look at some of the major calendars of the world.

A calendar can be based simply on the moon and consist of days and

The observatory at Jaipur, India, founded by Jai Singh in 1726.

7

lunar months, or it can be based simply on the sun and consist of days and solar years. Alternatively, some societies have tackled the difficult problem of basing the calendar on both the sun and the moon, while others have chosen to ignore both the sun and moon and to base their calendar on an arbitrary sequence of days. Let us look first at an example of a calendar based on the moon.

A calendar based on the moon: the Moslem calendar

The Moslem calendar is a surviving and important example of a purely lunar calendar. It is of comparatively recent origin, and records the Mohammedan era (Hegira) which started on the first day of the lunar month preceding the flight of Mohammed from Mecca to Medina in A.D. 622. One normal year of the Hegira consists of twelve months of alternate length 30 and 29 days which begin approximately with the new moon.

Since the average length of a lunar month is 29·53 days and the average length of the Moslem month is 29·50 days this calendar would get out of step with the moon by about 0·36 days (8 hours and 38 minutes) in one year and it would soon become noticeable that the months no longer started with the new moon. This is, of course, the basic problem of any lunar calendar and is comparatively simple to solve if an astronomer can tell you the true average length of a lunar month. For a satisfactory solution one needs to know the answer with surprising precision; in fact one needs to know the average length of a lunar month within a few seconds. The makers of the Moslem calendar came to this question fairly late in history and were spared the lengthy task of measuring the lunar month because it had already been done. For example, the Babylonians, who were expert and systematic astronomers, had already measured the lunar month with an error of less than 1 second about 1000 years before the flight of Mohammed. Given this information, the Moslems divided the calendar into cycles of 30 years; for the first 19 years of each cycle the twelfth month has its normal value of 29 days, but for the next 11 years it has 30 days. By this simple dodge the average length of the calendar month is increased from 29·50 to 29·530556 days which is about 0·000033 days or 3 seconds shorter than the true lunar month. Thus the Moslem calendar keeps the month in step with the moon so accurately that it would take about 2500 years for it to be wrong by one complete day.

However, the Moslem calendar, like all simple lunar calendars, does not keep step with the solar year and so the dates of the seasons drift slowly through the calendar. The average length of one year of the Hegira is about 11 days shorter than a solar year and so the dates in the Moslem calendar and in our Gregorian calendar, which is based on the solar year, coincide at intervals of about 33 years.

A purely lunar calendar in which the seasons have no fixed dates is unsuited to the everyday requirements of a modern state and can only continue to exist where it is supported by some powerful religious tradition. For this reason, almost all the early forms of lunar calendar have disappeared.

A calendar based on both the sun and the moon: the Babylonian calendar

The Babylonians tackled the difficult problem of basing a calendar on both the sun and the moon. They successfully incorporated the lunar month and the solar year into one system, the *lunisolar* year. They were not the only people to try this, for both the Greeks and the Romans, before Julius Caesar, had operated a sort of lunisolar calendar without much success. In particular, the Roman calendar was not systematically administered and got into such a hopeless muddle that it was, in effect, abandoned by Julius Caesar in 46 B.C. The Greek calendar was not much better. On the other hand the Babylonians, who had some expert mathematicians and astronomers, administered their complex lunisolar calendar systematically for several centuries. In a fairly advanced form their calendar survived the conquest of Babylon (539 B.C.) by the Persian kings and reached its final development under Persian rule close to 380 B.C. when it was modified to make use of the fact that 235 lunar months are equal to 6940 days or 19 solar years. This discovery is usually attributed to the Greek astronomer Meton in about 430 B.C., but modern research suggests that it was made before that date by the Babylonians. Meton may or may not have discovered it independently, nevertheless it is called the Metonic cycle. In fact it is a remarkably good approximation; 6940 days exceeds 19 solar years by about $9\frac{1}{2}$ hours and 235 lunar months by $7\frac{1}{2}$ hours. In practical terms this means that a workable lunisolar calendar can be designed if it is based on a cycle of 19 years, and this is what the Babylonians did.

The basic Babylonian year, like the Moslem year, consisted of 12 lunar months, each month starting with the new moon. As we have seen, such a year is about 11 days too short to keep in step with the solar year and so the Babylonians added an extra month in years 3, 6, 8, 11, 14, 17, and 19 of a 19-year cycle, thus making a Metonic cycle of 235 months in 19 solar years. Because some years had 12 months and some had 13, the dates of the seasons (equinoxes and solstices) varied from year to year by roughly 15 days or half a month. Presumably this variation was not sufficient to worry them unduly.

The Babylonian lunisolar calendar was one of the greatest triumphs of early astronomy and mathematics. I suppose that if history had been a little different we might have inherited it instead of the Roman solar calendar. As it was, their calendar survived for centuries in parts of the Near East and it is ironical that it was displaced, not so much by the Julian calendar of the Romans, but by the more primitive lunar calendar of the Moslem conquerors.

Calendars based on an arbitrary sequence of days: the Mayan sacred almanac and the astronomical calendar

In any account of the calendar the Mayans should be given a place. From looking at their monuments one gets the impression that they, more than any other society, were obsessed with the passage of time, rather in the same way as the ancient Egyptians were obsessed with life after death. Astronomers, or astronomer-priests, played an important part in Mayan affairs and they succeeded in making the calendar so complicated that they made themselves powerful and indispensable. The Mayans had three systems of reckoning the passage of time, a year of 365 days, an approximate year (tun) of 360 days for calculating large time intervals, and, most importantly, a sacred almanac of 260 days. The 260 days of the sacred almanac were designated by a sequence formed from all the possible combinations of 20 names with 13 numbers, each name and number representing a divinity. Every combination was believed to have a particular significance and influence on human affairs. The conduct of Mayan affairs appears to have been dominated by the almanac and yet its origin is uncertain.

The Mayan almanac is an interesting example of a calendar which is based on a religious or astrological idea and yet is not obviously tied to some easily observable astronomical event. At first sight it seems that this use of a simple sequence of days should have avoided all the problems of the solar year and the lunar month. But this was certainly not so because the Mayans were deeply concerned with astrology and divination and regarded the positions of the sun, moon, and planets, and particularly of Venus, as very significant. Astronomers were therefore needed to calculate these positions for dates in the sacred almanac and especially to predict events such as full moon, eclipses, etc. For this purpose they appear to have built observatories, such as the Caracol tower in Chichen Itza, and to have made a remarkably comprehensive and precise series of observations, particularly of the planets. As far as one can see, observational astronomy played a major part in Mayan affairs.

A second example of a calendar based on an arbitrary sequence of days is the calendar used internationally by astronomers. It is worth noting because it is the only calendar designed wholly for science. The days are simply numbered in sequence starting with day 0 which began at noon at

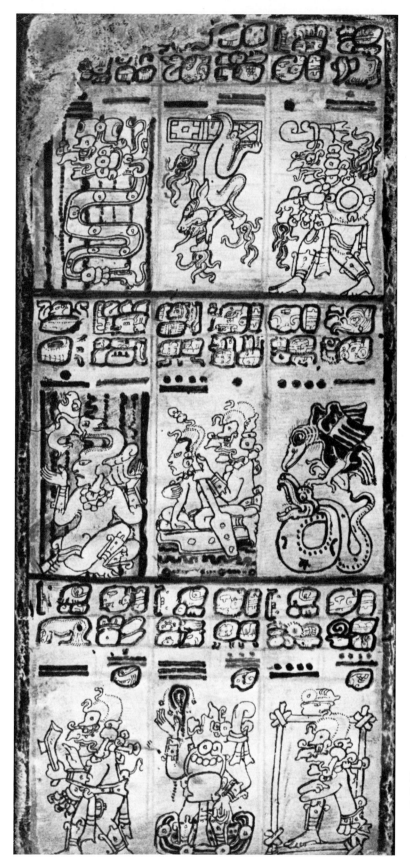

Part of the Mayan
sacred almanac
of 260 days.

Greenwich on 1 January 4713 B.C.! This delightfully archaic scheme was introduced in the 16th century by J. Scaliger who named the days of this calendar Julian days, after his father. This was a very confusing thing to do because the name has nothing to do with the Julian calendar which was named after Julius Caesar. The Julian day numbers increase in simple sequence and, as an illustration, the Julian day which lasts from noon on 31 December, 1975 to noon on 1 January 1976 (at Greenwich) is J.D. 2442778, which means that there have been 2 442 778 days since 1 January 4713 B.C.

At first sight the choice of 4713 B.C. looks like an attempt to fix day 0 at the Creation, like the Byzantine era used at Constantinople which dated from the supposed day of Creation, at that time believed to be 5509 B.C. But in fact the choice of 1 January 4713 B.C. has a less romantic and more technical origin. It is the most recent moment in time when three cycles, which in those days were regarded as important, were all in step. The first of these cycles is the so-called 28-year *solar cycle*; in the Roman Julian calendar the dates of each year occur on the same days of the week every 28 years. The second cycle is the 19-year *Metonic cycle* which we have already discussed. The third cycle is the 15-year *cycle of indiction* which was used to designate years in Roman times. This obscure origin of the start of the astronomical calendar has no practical significance—someone has to choose a day 0 and that is what Scaliger did. Astronomers are stuck with it.

A calendar based on the sun: the Gregorian calendar

The calendar which almost the whole world uses today, the Gregorian calendar, is directly descended from the Roman calendar introduced by Julius Caesar (the Julian calendar) in 46 B.C. For centuries the Romans had struggled with the awkward problems of reconciling the lunar month and solar year. They don't seem to have been as successful as the Babylonians, who not only knew how to regulate the calendar but actually did so systematically. The old Roman calendar year began in March and had 4 months with 31 days, 7 with 29 days, and 1 with 28 days. To keep in step with the solar year an extra month was added every other year with a length of 22 and 23 days alternately. This system gave an average year of $366\frac{1}{4}$ days, which is about one day too long, and so the seasons moved slowly through the calendar. However, this problem was not beyond the astronomers of the time who put forward an 8-year cycle which was considerably more accurate but, after about 200 B.C., became irrelevant because the administration of the calendar was no longer systematic and was probably exploited for political ends.

By the time of Julius Caesar the Roman calendar was in a muddle and had slipped by about 3 months with respect to the solar year. Caesar decided to abandon it and, with the help of the astronomer Sosigenes from Alexandria, he established a new calendar, the Julian calendar, in 46 B.C. This calendar represented a drastic break with tradition, it gave up all attempts to make a lunisolar calendar, and was based solely on the sun. Three years of 365 days were followed by one of 366 days (leap year), giving an average length of $365\frac{1}{4}$ days. Before its introduction the old calendar was brought into step with the seasons by increasing the length of the year 46 B.C. to 455 days. The new Julian calendar was then started on 1 January 45 B.C. Caesar ordained that January, March, May, July, September, and November should have 31 days and the other months should have 30 days, except February which should have 29 days in normal years and 30 days in leap years.

It is interesting, that although Julius Caesar employed a professional astronomer, he nevertheless made a calendar with the incorrect length of the solar year. The use of $365\frac{1}{4}$ days was not a sufficiently good approximation and if the best value for the solar year available at the time, from Alexandria or Greece, had been used it would have saved problems later on. A calendar year of $365\frac{1}{4}$ days is slightly longer than the true solar year and accumulates an error of 1 day in 128 years. That doesn't sound very much, but by about A.D. 1500 it amounted to about 10 days and had produced a noticeable effect on the date of Easter, which is a cardinal point in the Church's calendar. Since the days when Christianity was made the official religion of the Roman Empire the Church had controlled the calendar and so we find Pope Sixtus IV in 1474 inviting Regiomontanus, the leading astronomer of the day, to superintend the reformation of the Julian calendar. However, the death of Regiomontanus stopped the work and nothing further was done for many years until Ghiraldi, a physician and astronomer of Naples, proposed a new calendar. This calendar was finally introduced by Pope Gregory XIII in 1582 and the man who contributed most to it was Christopher Kley (better known as Clavius) who made the calculations and explained the whole complicated business in an astonishingly comprehensive book (*Romani Calendarii a Gregorio XIII*, Rome 1603). Clavius was a well-known mathematician and astronomer and is perhaps better known for his contributions to geometry; he has been called 'the Euclid of the 16th century'. To wipe out the accumulated error of the Julian calendar, 10 days were omitted from October 1582, not without some disquiet. The rule for leap year was altered to omit the extra day in February in all centurial years that are not divisible by 400 (e.g. 1700, 1800, 1900, 2100). By means of this simple scheme the average length of a calendar year was decreased from 365·2500 to 365·2425, which is very close to the true value 365·2422. The error in the Gregorian calendar is about 26 seconds in one year and will amount to one complete day in about 3300 years, so we have plenty of time to think about it.

When it was introduced by Gregory XIII in 1582 the Gregorian calendar was immediately adopted by Roman Catholic countries. However, in

A detail from Hogarth's *Humours of an Election Entertainment*. The banner, 'Give us our Eleven Days', refers to the resentment felt over the introduction of the Gregorian calendar in 1752.

Protestant countries it took much longer and it was not introduced into Great Britain until 1752. An Act of Parliament passed in 1751 decreed that 2 September 1752 should become 14 September 1752 and the year 1752 should begin on 1 January and not, as before, on 25 March. The omission of 11 days from the calendar was, of course, necessary to correct the accumulated error of the Julian calendar and caused very considerable trouble and, in fact, some riots. Many people felt that they had been robbed of these 11 days by the Government and strongly resented it. The change in the starting date of the year from 25 March to 1 January sounds rather startling but one must realize that many different starting dates for the old Julian calendar were in use. For example, in England before the 14th Century, the Julian calendar began on 25 December with the Nativity. It was then moved to the feast of the Annunciation (25 March) and finally back to the Circumcision (1 January). The Scots had used 1 January for the start of their year since A.D. 1600.

14

There is one curious point about starting dates which is rather interesting. The Christian era starts with A.D. 1 and the year before that is 1 B.C. There is no year A.D. 0 nor 0 B.C. This arrangement gives rise to endless arguments at the end of each century as to when the next century begins. Thus, contrary to most peoples' expectations and also I would say, to common sense, the 20th century will end on 31 December, A.D. 2000. We may feel like celebrating the start of the 21st century on 1 January, A.D. 2000 but, strictly speaking, we should wait until 1 January, 2001.

When Julius Caesar reformed the Roman calendar in 46 B.C. we lost touch with the moon as a convenient natural calendar and it now wanders like a lost soul through our year and we have to consult almanacs to find out when it will be full and when it will be new. Nevertheless, although deserted, the moon left her mark, the month, on our calendar as a reminder of more primitive times. Our month is, as we have already seen, about one day longer than a true lunar month so that the dates of the moon advance by about 11 days each year. In earlier times, especially before street lighting, people took a greater interest in the phases of the moon, as one can see from those beautiful clocks of the 18th and 19th centuries which show the phases of moon together with the date. Nowadays I suppose that interest in the phases of the moon has become specialized; they concern some country dwellers, astronomers, people planning military operations and burglaries, and tourist agencies planning visits to the Taj Mahal! Anyway, there is still sufficient interest to warrant the inclusion of tables of the phases of the moon in almost every pocket diary.

The ecclesiastical calendar

The adoption of our purely solar calendar solved, almost but not quite, the problems of the civil authorities. The main purpose of the calendar was to fix the dates of the seasons, which it does very well. Thus, the dates of the equinoxes and of mid-summer and mid-winter move very little from year to year; they move, about 1 day either side of their average values, very slowly over a period of about 200 years. For example, the date of mid-summer moves between 22 June and 20 June, while the vernal equinox moves between 21 March and 19 March. However, one minor thorn in the side of the rationally-minded citizen is the variable date of Easter. From the point of view of a modern secular administration it is absurd that the date of one of the major holidays of the year should be determined by an ecclestiastical calendar in which the date may vary from 22 March to 25 April. Furthermore, there is no obvious sequence in the dates of Easter. In the Julian calendar they were repeated every 400 years but in the Gregorian calendar they no longer repeat at any reasonable interval, in fact a simple-minded calculation shows that they repeat every 5 700 000 years!

15

The face of an English astronomical long-case clock, with moon dial, by Benson, 1760.

The ecclesiastical calendar is a wonderful mixture of the Hebrew lunar calendar, with its Metonic cycle, and our Gregorian solar calendar. The moon fixes the movable feasts and the sun the immovable feasts. For example, Christmas Day is celebrated on a fixed date in the Gregorian calendar and is an immovable feast. On the other hand, Easter Day is movable in the Gregorian calendar and all the other movable feasts, Rogation, Ascension, etc. follow in a fixed sequence. The *Book of Common Prayer* tells us how to find the date of Easter; 'it is always the first Sunday after the full moon which happened upon or is next after the twenty-first day of March; and if the full moon happens upon a Sunday, Easter Day is the Sunday after'. To help us with this calculation there are tables in the prayer book which involve the beautiful and mysterious words to which we referred earlier. The golden number identifies the number of the year in the Metonic cycle of 19 years and is of Greek origin, the dominical letter identifies the dates of Sundays in the year and is of Roman origin, the full moon is the paschal full moon and is of Hebrew origin.

All this sounds fairly simple, provided that one doesn't look too closely at the fine print. In fact, when the Gregorian calendar was introduced, much fuss was made over the date of Easter and it was decided to make it crystal clear when Easter should be celebrated for years to come. The rules were laid down so that no further astronomical observations would ever be needed. To this end a set of tables for finding the age of the moon were calculated from the Metonic cycle and the vernal equinox was fixed at 21 March. It was also decided that the date of Easter should be the same anywhere in the world. But from an astronomer's point of view these rules are not straightforward. For example it is amusing to note that the moon which is used to fix the date of Easter is an 'ecclesiastical moon' defined by the tables, and is not precisely the real moon. Also the vernal equinox is taken as 21 March which is not necessarily the true equinox for a particular year, nor was it the true equinox at the Council of Nicea where these rules were drawn up. As a consequence, the date of Easter, as calculated from the official rules can differ very occasionally, and only in some parts of the world, from the date one would arrive at by looking at the sky. It is not surprising that at the Congress of the Eastern Orthodox Churches in 1923 it was decided not to follow these rules but to determine the date of Easter by the real moon at the meridian of Jerusalem.

Conclusion

In this brief review we have seen that the making of our calendar was not as simple as it looks. The three natural clocks in the sky, the day, the lunar month, and the solar year are awkwardly related and it was a great achievement of observational astronomy to measure these relations with sufficient precision to make a satisfactory calendar. Our modern Gregorian calendar which looks so deceptively simple is in fact based on a remarkably precise knowledge of the periodicities of the solar system. One must know the length of the solar year with an accuracy of about one day in 3000 years. Furthermore, the calendar is based on a bold decision not to try to reconcile the sun and the moon in a lunisolar calendar, and this decision itself is based on centuries of experience.

Looking back, the precision of some of the results obtained by early astronomers before the invention of the telescope, the micrometer eye-piece, and the accurate clock, is most impressive. All the measurements were made with the naked eye by sighting along stone monuments, lining up distant stones or, in some cases, with instruments using some simple form of sighting device. By modern standards these observations were coarse, but some of them, nevertheless, yielded remarkably precise results, not for astronomical positions, but for periodicities, because they had been made systematically over long periods of time.

Consider, for example, the length of the lunar month. The Mayans, in calculations recorded in the 7th century A.D., took 405 moons to be equal to 46 sacred 260-day almanacs. We now know that the error in this equation is between 2 and 3 hours in 33 years. If their method of measuring the lunar month was uncertain to one day, then it would have taken at least 300 years for the Mayans to reach the precision which they apparently did.

Lengthy astronomical records were used, not only to measure obvious periodicities such as the month and year, but to recognize far more obscure patterns of events. An outstanding example is the remarkable discovery made by the Chaldeans that the sequence of eclipses of the sun and moon is repeated every 6585 days or 223 lunar months. Edmond Halley, in 1691, erroneously called this cycle the *saros* and it has been known by that name ever since. When the saros was originally discovered it must have been a real breakthrough in the astrological world because eclipses were regarded as peculiarly portentous. It must have greatly enhanced the prestige and perhaps, also the income of the Chaldean astronomer-priests.

Looking at the physical remains of the early observatories or observatory-temples at Stonehenge or Carnac, in Egypt or Mexico, one gets a vivid impression of the importance which some early societies attached to watching the sky. There are, unfortunately, no records of the observations made at Stonehenge but it appears to have been an observatory, or observatory-temple, built in about 2000 B.C. and one of its principal functions must have been to observe the solstices, mid-summer and mid-winter. It was literally a calendar in stone. There are many other ancient observatories which differ in detail, according to the local interests. The Egyptians, for example, constructed some of their buildings to observe the rising of Sirius with the sun, while Mayans did the same for the planet Venus. The motivation to build these observatories, or observatory-

Stonehenge—probably an early astronomical observatory–temple dating from 2000 B.C. The outer circle of 56 white dots are the Aubrey stones.

temples, must have been very strong and presumably was largely religious, concerned with omens, prophecy, and divination but, at the same time, there was always a practical agricultural motive for regulating the calendar. It seems unlikely, however, that any simple utilitarian motive can explain Stonehenge. The big stones, sarsen stones, are said to weigh 50 tons and to have required 800 men to move them several miles. In modern terms the building of Stonehenge must have taken more of the gross national product than sending a man to the moon in our own day, and I suspect that the motives for both were equally irrational.

The remains of a neolithic observatory at Carnac in north-western France.

To modern astronomers the whole topic of calendars is dead and buried. All that is left to do now is for a few experts to keep track of the solar system in terms of the calendar and publish their results annually in the astronomical ephemeris. Nevertheless, although the making of the calendar is a finished chapter in the history of science, it is an important chapter. Not only is the final product, the calendar, of practical value to society but the systematic observation and recording of the astronomical events on which it is based, was itself a significant step in the development of science. All the periodicities in the solar system, the Metonic cycle, and the saros, were found by observation before there was any theory of celestial mechanics. Their discovery demonstrated that there are unsuspected regularities in nature and that these regularities can be discovered by patient observation of what actually takes place in front of our eyes. This is an important step away from a view of a physical world controlled by a host of unseen angelic and demonic forces, towards a view of a world in which systematic observation and reason lead to deeper understanding and control.

2 Stars and Time

It was only recently, while my wrist-watch was being repaired, that I came to realize how widely it is assumed that the good citizen knows the time, at least to the nearest five minutes. The law, for example, takes it for granted that somehow or other, the law-abiding motorist knows the time. The regulations about parking, freeways, and one-way streets are all written up on the assumption that we know the time even where there is no public clock. And the same is true of almost everything else, concerts, theatres, and even going to Church, where in many cities they no longer ring the bell to tell us when to come. One has only to tune in to a breakfast programme on the radio to be reminded how anxious people are to know the time. If, for a moment, we could turn the clock back to what our history books call the Middle Ages we should, I believe, find a great difference. They were not, like us, slaves to the clock. Indeed, before the 18th century there were very few clocks in private homes; most of them were in monasteries, churches, or palaces. I suppose that in those days they thought in coarser divisions of the day, like early morning or late afternoon, and didn't fret so much as we do about being kept waiting half an hour for an appointment.

Apart from these historical differences in the use of time, there are more interesting differences in the way people thought about time. Press someone today to describe time and they will probably talk about a uniform flow of successive instants, an imaginary straight line along which the moment 'now' moves at a steady speed from an infinite past into an infinite future. But in earlier days you might have been offered a different answer. Perhaps the most common view pictured time as being cyclic, a repetition of events on the stage of a permanent, unchanging, and eternal universe. Hindus still see time in this way, as an endless cycle of death and rebirth. To them, time is not something precious to be saved, rather it is the soul which is precious and has to be saved by freeing it from this cycle. If, on the other hand, we look at the Christian Church, remembering that for centuries it was the curator of our timekeeping and calendar, we find a different view. Here, the dominant theme is expressed in the words of a well-known hymn, 'God is working his purpose out as year succeeds to year'. Time is regarded as a stage on which God unfolds his plan from Creation to Judgement. There is only one performance of the play, and after it is finished, the programme calls for a timeless eternity of Heaven and Hell. Again like the Hindu, the

Astronomical clock at Hampton Court Palace, made for Henry VIII in 1540.

Christian sees the perfect state as timeless; evidently, whatever time may be, it is something that many people would like to escape.

What, then, is time? Most of us feel that we know the answer perfectly well, but find it hard to put into words. We end up either with a tangled web of words or with a brief tautology. Even St Augustine felt obliged to preface a lengthy discussion of the nature of time with the words, 'What then is time? If no one asks me, I know. If I want to explain it to a questioner, I do not know'. Of course, he doesn't leave the question there but goes on to discuss the nature of time in page after page of fruitless metaphysics so beloved by early scholars. The trouble is, not that St Augustine couldn't think clearly, but that he was asking the wrong sort of question. Experience shows that a question like 'What is time?' is not likely to advance our understanding of the physical world. Maybe it is a question which we should put to an artist who, in painting or music, might capture some of that elusive knowledge which we feel we have, but cannot put into words. But it is not the right question to ask a scientist. Broadly speaking, progress in understanding the physical world has depended on asking the right questions, not grand questions like 'What is life, or mind, or time?', but modest, bite-sized questions, which can be answered by observation, experiment, or calculation. Most of the bricks in the imposing façade of modern science are, if you look closely, quite small. As a general guide to the right sort of question, experience suggests that it is better to start with *how*, not why or what. Let us then follow this prescription and ask 'How is time measured?'.

The most ancient answer to this question is that time has always been measured by counting days. The clock was the sun and it measured the three most important events of the day, sunrise, noon, and sunset. But the need to subdivide the day still further must have arisen early in history and so man invented the shadow clock. The shadow of the sun cast by a rock, post, or obelisk fell on a graduated scale, such as a flight of steps, and people told the time by counting the steps. One of the earliest references to a shadow clock of this type is the promise of a sign given by the Lord to Hezekiah (Isa. 38), 'Behold I will make the shadow cast by the declining sun on the dial of Ahaz turn back ten steps'. An interesting feature of that story is that the Lord gave Hezekiah the choice of seeing the shadow turn forward

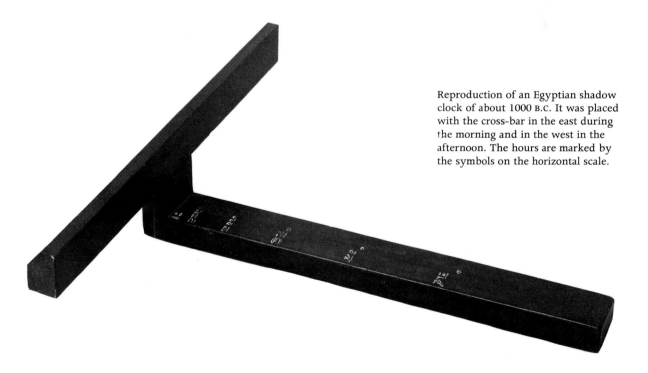

Reproduction of an Egyptian shadow clock of about 1000 B.C. It was placed with the cross-bar in the east during the morning and in the west in the afternoon. The hours are marked by the symbols on the horizontal scale.

or back and Hezekiah very firmly chose that it should go back (2 Kgs, 20). He felt that a backward movement would be more convincing; after all the shadow would go forward naturally in the normal course of the day.

The story of Hezekiah dates from about 700 B.C. and since then there has been a good deal of effort put into measuring time. Indeed one of the foundations of our civilization is a sound system of measuring time by the sun. The problems of developing this system proved to be far from trivial and their solution is one of the major contributions which the study of astronomy has made to society. Let us take a brief look at what these problems were.

Measuring time by the sun

The most obvious difficulty in subdividing the day by measuring shadows is that the path of the sun across the sky changes from day to day. This is due to the fact that the Earth rotates about an axis which is inclined at $23\frac{1}{2}°$ to the plane of its orbit around the sun. Thus at mid-summer in the northern hemisphere the Earth is at a position where the North Pole is tilted towards the sun; Londoners see the sun 62° above the horizon at noon, the shadows are short and the days are long. At mid-winter the North Pole is tipped away from the sun which appears low in the sky (15° for Londoners at noon) giving long shadows and short days. Furthermore, the direction as well as the length of the shadows vary; for example, the direction of the sun at sunrise and sunset varies from summer to winter by an amount which depends upon latitude. It is these variations in the length and direction of the sun's shadow that make it difficult to design a shadow clock with a simple scale or dial which can be used throughout the whole year.

The first question to decide is whether or not to distinguish clearly between day and night. If we prefer, we can subdivide the whole of one rotation of the Earth (day plus night) into equal parts. On the other hand we can treat day and night as two completely separate components of the calendar and subdivide them into units of daylight and units of darkness. When one considers the profound difference between day and night, especially in primitive experience, it is not surprising that most early systems of the timekeeping were based on the second alternative; they chose to regard day and night as quite separate events with their own distinct measures. This was certainly so in Europe as late as the 14th century and is perhaps the reason why we have no word for 'day plus night' in the English language. The Saxons divided daylight into four tides, as we are reminded by the words noontide and eventide, but the more general practice was to divide the time between sunrise and sunset into a fixed number of 'seasonal' or 'unequal' hours usually 12 hours of daylight and 12 hours of night. As a consequence the length of an 'unequal' hour

24

A 17th-century column sundial. The dragon is turned until its tail is over the appropriate month as shown by the signs of the zodiac. The shadow of the tail then points to the time on a scale marked on the column.

A portable sundial built into a square box 25 cm long, and made of gilded brass. On the top, outside, is a map of the northern hemisphere and on the bottom, one of the southern hemisphere. The cover has a small aperture that is opened and closed by a disc pivoted on the inside of the cover. This disc shows the position of the sun against a scale corresponding to the days of the lunar month. Inside the box there is an arc graduated for altitudes and an equinoctial circle for solar hours. A wind rose and magnetic compass are in the base.

depended on both season and latitude. In London a daylight hour was roughly twice as long at mid-summer as at mid-winter, while at the equator the hours were constant in length throughout the year.

This seasonal variation in the length of the hour is the first complication which we, as astronomers, face in marking out a scale of time for our sundial. No matter how we design this sundial, we must inevitably draw different scales for different seasons and the number of scales will depend on how accurately we need to tell the time. In early sundials there were three scales, two for the solstices (mid-summer and mid-winter) and one for the equinoxes (spring and autumn). Later sundials were much more complicated and were often marked with a separate scale of hours for each month.

For many purposes these unequal hours made good sense; after all, if a job had to be done in daylight then there is less time to do it in winter than in summer, and it seems reasonable to make the hours shorter. However, unequal hours did not suit the needs of organized society which came to depend more and more on people knowing the time fairly accurately.

Sundial at Merton College, Oxford. The shadow of the ball shows local solar time on the scale of Roman ▷ numerals; on the other scale the sign of the zodiac corresponding to the declination of the sun. Designed in 1974, it is in the same position as an 18th century dial.

26

A Japanese clock with two escapements—one for day and one for night.

Hours of variable length were obviously incompatible with mechanical clocks running at a uniform rate. One has only to look at some early European clocks or, better still at Japanese clocks, to realize how true that was. Many European clocks tackled the difficult problem of showing unequal hours by having several scales of hours, like early sundials, but the Japanese made their clocks run at two different rates, one by day and one by night. They often had two separate escapements which were connected alternately to the driving train after 12 hours; the rate of these escapements could be adjusted independently according to the season of the year by sliding weights along a graduated bar.

The use of unequal hours appears to have been abandoned in China at about the time of the Han dynasty (2nd century B.C.); in contrast Japan continued to use unequal hours until 1873. In Europe unequal hours fell into disuse in the 14th and 15th centuries and an alternative system of 'equal hours' or 'equinoctial hours', already in use by astronomers, was slowly adopted. The equal hours were defined by dividing one complete rotation of the Earth, a solar day, into 24 equal parts.

For a time in Europe equal and unequal hours were both in use and there are some interesting sundials and mechanical clocks which were designed to show both systems at once. But the use of equal hours eventually prevailed and greatly simplified the design of all forms of timekeeper, including the sundial. Our familiar garden sundial is not, as many people believe, a replica of some archaic device but is of comparatively modern design dating from the period when Europe adopted equal hours. It is much simpler than earlier sundials, having only one scale of hours for the whole year. This simplification was a great advance and was made possible by two things; the introduction of equal hours and the use of an inclined post or gnomon to cast the shadow. Unless one is prepared to live at the North or South Pole one cannot make a simple sundial with a vertical gnomon. This is because it is only at noon, when the sun is due south (or north) that the shadow of a vertical post is always in the same direction throughout the year. At any other time of day, the direction of the shadow is not always the same because, as we have seen, the path of the sun across the sky changes with season due to the tilt of the Earth's axis. Therefore, even if we adopt equal hours, we must still mark out different scales of time around a vertical gnomon for different times of the year. There is one way in which this can be avoided and that is to move the gnomon or the scale. Thus it is possible to design a sundial with a single scale of hours if provision is made to move the gnomon to different positions along a north–south line according to the time of year. Sundials made in this way are rare and are known as analemmatic sundials; the most famous example in Europe is in the grounds of the cathedral at Bourg-en-Bresse in France.

There is, however, a complete solution to this problem which is incorporated in our garden sundial. The post or gnomon is not vertical but is mounted parallel to the axis about which the Earth spins. It is therefore aligned north and south and is inclined at an angle to the horizontal equal to the latitude; the gnomon would be horizontal at the equator and only vertical at the North or South Pole. This brilliant idea ensures that as the Earth rotates the sun appears to make a circle around the gnomon every 24 hours. Thus, if we mark the direction of the shadow at noon, the direction at any other time will measure the angle through which the Earth has rotated since noon and hence will be a direct measure of the time. For a given time this angle, and hence the direction of the shadow, will always be the same; it will not depend, as it does with a vertical post, on the season. It is therefore possible, by using an inclined gnomon, to mark out one single scale of hours for the whole year. This scale can be marked out on any surface we like, horizontal, vertical, or at any other angle, and the only remaining problem is that the hourly divisions are not uniform in length. The only way to solve that difficulty is to engrave the scale on a dial which is parallel to the Earth's equator so that the hour marks are equally spaced

15° apart. Sundials made in that way are called 'equatorial dials' and have been used in China for centuries.

Mean solar time

Following the introduction of equal hours in the 14th and 15th centuries, a second and far less obvious problem in keeping time by the sun emerged as mechanical clocks improved. The early clocks had a verge and foliot escapement and usually only one hand; most domestic clocks were only capable of keeping time to the nearest quarter of an hour and were often

A modern equatorial sundial designed by Henry Moore. Made of bronze, it is pictured here in Printing House Square, London. It is now at La Hulpe in Belgium.

An early wooden-frame chamber clock made in Germany in 1643. Clocks like this were introduced into domestic houses towards the end of the 14th century. They were usually hung in the hall so that their striking could be heard in the other 'chambers'. The rate of the clock is controlled by the horizontal swinging bar (the foliot) and can be adjusted by moving the hanging weight along the bar.

checked against a sundial. However, the introduction of the pendulum in about 1660 made the domestic clock much more accurate and, for the first time, many households were able to tell the time by day and by night to the nearest minute. This was the stage at which the sundial really started to become obsolete and it is interesting to note that some clocks of this period were made from parts originally intended for sundials; brass was in those days, as now, an expensive metal. With the advent of better clocks it became apparent that the system of civil timekeeping had to take account of the fact that the length of the solar day varies throughout the year.

It has been known, certainly since the days of Ptolemy (A.D. 150), that the length of the solar day, measured from noon to noon, is not constant. It depends on two factors, the rotation of the Earth about its axis and the orbital motion of the Earth around the sun. We may assume, for the moment, that the rotation of the Earth is perfectly steady and that any variations in the length of the solar day are due to its orbital motion. Because the Earth's axis is tilted, the sun does not appear to move steadily round the celestial equator in the course of the year, but moves along a path called the ecliptic, which weaves about the equator. As a consequence the distance which the sun travels along the celestial equator is not the same from one noon to the next and so the length of the solar day varies throughout the year.

A second cause of variation in the solar day is that the Earth's orbit around the sun is an ellipse, not a circle. Consequently both the speed of the Earth in its orbit and its distance from the sun vary throughout the year. This distance varies by about 3 per cent and it is amusing to note that, contrary to what many of us think, the Earth is farthest from the sun during the northern summer and nearest during the northern winter; in fact the sun actually looks larger in the winter. These changes in the speed of the Earth, together with the inclination of the ecliptic to the equator, cause the length of the solar day, measured from noon to noon, to vary throughout the year by nearly one minute. And so, if we want to keep time to the nearest minute with a mechanical clock which runs at a uniform rate, then we cannot base our system of timekeeping directly on the solar day.

This tricky problem had already been solved by astronomers. They invented an imaginary body called the *mean sun* which moves at a perfectly uniform rate along the celestial equator, not along the ecliptic. Its position on the celestial equator (right ascension) is defined by a formula which makes it precisely equal to that of the real sun at the equinoxes, and keeps it as close as possible to the position of the real sun during the rest of the year; in this way the length of the year is still controlled by the real sun. Sometimes the mean sun runs ahead of the real sun and sometimes behind. *Mean solar time* is defined with respect to the mean sun and, because its motion along the celestial equator is perfectly uniform, the mean solar day, hour, minute, and second are perfectly uniform in length throughout the year.

There is, of course a difference between the time with respect to the mean sun and the time with respect to the real sun. This difference is given by the equation of time:

mean solar time − apparent solar time = equation of time

and varies either way by about quarter of an hour throughout the year. If, therefore, we wish to check a clock, showing mean solar time, by a sundial, showing apparent solar time, we must know the appropriate value of the equation of time for the day. One way of finding this information is to look at a terrestrial globe; there is usually an elongated figure of eight, called an *analemma*, printed on it somewhere, often in the middle of the Pacific Ocean. This strange figure is a graph showing how the equation of time varies with the time of year.

Measuring time by the stars

We have so far considered only one way of measuring time, by the sun. And yet there is another way, perhaps just as ancient, and that is to measure time by the stars. As the Earth spins on its axis the whole sky appears to turn and

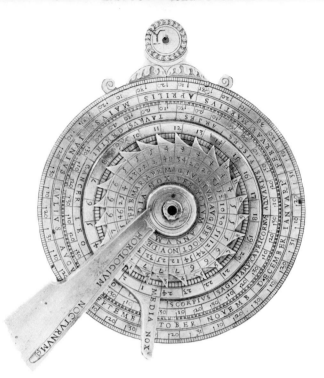

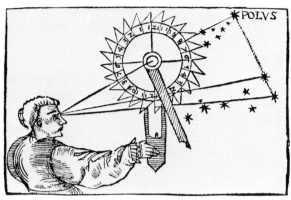

Left A night dial or nocturnal, made in 1568 for the latitude of Florence. The outer brass disc is 145 mm in diameter and includes a zodiacal calendar. The toothed disc is inscribed with information about the length of the night at various times of the year.
Above Using a nocturnal to determine time at night by the stars. From a book published in 1539.

if one watches a particular star it circles the celestial pole like the hour hand of a clock. Therefore a simple way of measuring time is to hold a clock face up to the sky with its centre on the pole star, move the hour hand to point at the star for which the instrument is designed and read the time off the dial. Such an instrument was in fact used in the late Middle Ages and was called a night dial or *nocturnal*.

Time roughly measured in this way is called *sidereal time* and is of special importance to navigators, astronomers, surveyors or anyone concerned with making measurements with respect to the spinning Earth. Basically, it is much simpler than solar time because it depends solely on the rotation of the Earth on its axis and not on its orbital motion around the sun. Sidereal days, hours, minutes, and seconds are therefore uniform in length throughout the year. The start of each sidereal day is defined, not in terms of a bright star as one might imagine, but as the moment—sidereal noon—when a point in the sky, called the first point of Aries, crosses the meridian. The first point of Aries is the position of the sun on the celestial equator at the vernal equinox.

A sidereal day is shorter than the solar day because the Earth travels once round the sun in a year and so makes one less revolution with respect to the sun than the stars. Roughly speaking, 365 solar days are equal to 366 sidereal days and so the sidereal day is about 4 minutes shorter than the solar day and the sidereal hour is 10 seconds shorter than the solar hour. There is another awkward difference between the two systems of time. The sidereal day is defined to start at sidereal noon while the solar day starts at solar midnight. Because of this, on 20 March at the vernal equinox, a solar and a sidereal clock show times which are exactly 12 hours apart; six months later, in September, they show the same time.

Thus, although sidereal time is basically simpler than solar time, it is not suitable for everyday life because it does not keep in step with the sun. No one wants to live in a world where 12 o'clock is sometimes at noon and six months later has moved round to the middle of the night.

The quest for precision

The history of time-keeping is largely one of an apparently insatiable demand for higher and higher precision. At the beginning of the 18th century a pendulum clock might have kept time with an accuracy of several seconds per day, but this was soon to be improved by the introduction of better escapements, by compensation for changes in the length of the pendulum with temperature, and by providing a source of power to keep the clock going while it was being wound up. So effective were these improvements that by 1889 the Riefler clock, widely used by astronomers, was capable of keeping time to better than $\frac{1}{10}$ second per day. The final stage in the development was reached in 1921 with the introduction of the Shortt clock which could keep time to better than $\frac{1}{100}$ second per day. Thus in about 220 years the accuracy of the pendulum clock improved by more than 1000 times; nevertheless, like the sundial, it was displaced by something more precise.

Its successor, the quartz clock, was the product of that precocious child of the twentieth century, the electronic engineer, who revolutionized timekeeping and made the use of mechanical clocks for very precise timekeeping completely obsolete. The heart of this new type of clock is a small block of quartz which is made to vibrate, rather like a bell, at its natural frequency; this frequency, like the note of a bell, depends on its size. These vibrations are counted and their sum is the 'time' shown by the clock. When it was first introduced as the standard timekeeper at Greenwich in 1942, the quartz clock was about ten times better than the best pendulum clock and would keep time to about $\frac{1}{1000}$ second (1 millisecond) per day; more recent quartz clocks are very much better and are capable of running so regularly that they vary between $\frac{1}{10}$ and $\frac{1}{1000}$ milliseconds per day.

Despite its remarkable performance, however the quartz clock had a short life as a standard of time and was displaced by the atomic clock in 1967. The standard atomic clock counts the vibrations of one particular type of atom, the caesium atom; there are other atomic clocks based on other atoms, for example on rubidium or hydrogen, but the caesium clock is now the preferred standard. Atomic clocks are capable of running with incredible regularity, but they have another important advantage; they provide a standard of time interval which depends on the structure of the

34

atom and can therefore be reproduced in any laboratory in the world because, as far as we know, all atoms of the same element are identical. By comparison the rate of a quartz clock depends upon the precision with which a block of quartz can be made and can only be 'standardized' by comparison with another standard clock. At the present time atomic clocks are in an early stage of development but they are already capable of running with a regularity which is better than 1 microsecond (1 millionth of a second) per day corresponding to a variation of 1 second in about 3000 years. Even this not the limit, more precise clocks are foreshadowed and one cannot foresee where the search for precision will end; sometimes one wonders if it is some form of obsession!

This remarkable advance in the precision of clocks is a feature of the rapid development of electronic engineering which is still taking place. The widespread application of electronics to communication, radio, television, navigation, surveying, radio-astronomy, and to a wide variety of other physical measurements has brought with it the requirement for very precise standards of frequency. Since the frequency of any phenomenon is simply a count of how often it occurs in a given time interval, a precise standard of frequency is necessarily also a precise standard of time interval and vice versa. Thus it was really the demand by engineers and physicists for better standards of frequency which led to the demand for better standards of time and gave practical importance and urgency to a new question which now faced astronomers, 'Is the Earth a good standard clock?'

The earth as a standard clock

To answer this question we must go back to 1695 when the second Astronomer Royal, Edmond Halley, drew attention to the fact that the moon appears to be accelerating in its orbit around the Earth. The effect which he reported is exceedingly small, an advance in the position of the moon of about 10 seconds of arc per century and it is remarkable that Halley discovered it. In fact Halley found the effect by calculating the times of ancient eclipses from contemporary observations of the motion of the moon on the assumption that the rotation and orbital motions of the Earth are perfectly constant. A comparison of his calculations with ancient records of eclipses showed that, if the Earth is constant, then the moon must be accelerating. There seemed at first to be no plausible explanation for the effect, and this was a shock to a generation of astronomers brought up in the long shadow of Newton to believe that all the motions of the solar system could be explained. However, despite attempts by men of great ability, like Euler and Lagrange, it was not until 1787 that the great mathematician Laplace offered an explanation of the moon's acceleration in terms of gravitational theory. He showed that the gravitational pull of the other planets must slowly modify the Earth's orbit around the sun and that, as a consequence, the effect of the sun's pull on the moon decreases and the moon must therefore accelerate in its orbit. This explanation was accepted for some years until J. C. Adams—the man who predicted the position of Neptune before it was discovered—showed that Laplace's theory only accounted for half of the observed acceleration. In the following years a great deal of attention was paid to this problem by astronomers and the most plausible explanation which emerged was that the unexplained fraction of the moon's acceleration is due to the slowing down of the Earth. The ocean tides dissipate energy in friction and slow down the spin of the Earth thereby making it appear that the moon is accelerating. At first this theory looked very promising but it soon ran into difficulties when it was calculated that the loss of energy in the tides of the main oceans of the world is much too small to explain the effect. It was later suggested that the majority of tidal friction occurs in shallow waters, not in the main oceans, and it was claimed that the slowing down of the Earth could still be explained in this way. However, in this explanation, no less than three quarters of the total loss of energy is supposed to take place in the eastern Bering Sea and this estimate has recently been questioned on the grounds that the tidal current in that region has been seriously overestimated. There are also other factors which may contribute significantly; for example, there are tides in the solid rocks of the Earth's mantle as well as in the oceans and these Earth tides must dissipate energy; also there is the possibility that the magnetic field of the Earth may slow down its rotation by interacting with the surrounding electrified regions called the magnetosphere. It has even been suggested that the effect may be partly due to a progressive decrease in the force of gravitation. All these effects, like the ocean tides, are hard to assess quantitatively. In the meantime we must accept that, probably for a variety of reasons, the spin of the Earth is slowing down and the length of the day is increasing by about 1 millisecond in the course of a century. This doesn't sound very much but it means that the Earth 'loses time', compared with a perfectly uniform clock, by about 20 seconds per century.

In reviewing the performance of the Earth as a clock there is something else to be taken into account. Not only is the spin slowing down but it is irregular. The first evidence for this was put forward by the astronomer Simon Newcomb in 1870. By analysing observations of the moon as far back as 1675 he was able to show that its apparent motion is not varying smoothly but is irregular. Similar irregularities were observed in the apparent motions of Mercury, Venus, and the sun and so it was reasonable to conclude that these irregularities are not in the motion of the moon but in the rotation of the Earth.

Since the advent of very precise clocks it has been possible to confirm the slowing down of the Earth and of the irregularities pointed out by Newcomb; furthermore, these clocks have shown that there are seasonal variations in the rate of rotation. The irregularities amount to a change in the length of the day of as much as 5 milliseconds over a period of a year or so; the seasonal variations are much smaller, about 1 millisecond. The reasons for the irregular changes are still not understood and a variety of possible causes, such as variations in vegetation, sea level, ground water, ice, and snow have been explored without success. One possible explanation remains, that they are due to changes deep down in the core of the Earth. Attempts to explain the seasonal effects have been more successful. Calculations suggest that most of the observed effects can be explained by the seasonal pattern of winds.

At this point one may well ask how the performance of a clock can be measured if we claim that it is superior to our basic standard of time, the rotation of the Earth. The answer is that it can only be done if we have another clock, a similar clock or preferably a superior clock, and if we are prepared to make some assumptions. In the case of the Earth, Halley used data about eclipses to compare its rate of rotation with the orbital motions of the solar system; thus when he concluded from these data that it is the Earth that is slowing down, he was assuming that the whole solar system is a superior clock to the Earth. In principle we can measure the performance of any other clock in the same way by comparing it with the orbital motions of the solar system; but, in practice, this is not a convenient method of testing clocks because it takes so long to establish the time with the necessary accuracy. In practice, most of the work on the stability of very precise clocks is done by inter-comparing the rates of several clocks under rigorously controlled and very stable conditions. The assumption is usually made that their average rates are uniform and any changes in their relative rates are taken as a measure of their stability. This assumption is not always justified and, as an example, the effects of ageing in quartz clocks proved to be difficult to establish by comparing different clocks because they all tended to age together.

Returning now to our earlier question, 'Is the Earth a good standard clock?', it is clear that for everyday life the answer is yes, but for certain specialized scientific and technical purposes the answer is no. For everyday life where we are hardly ever concerned to measure time to better than a second or two, the Earth is a perfectly satisfactory clock; we are content with a system of time which keeps in step with the sun and minor irregularities don't worry us. But for many more specialized purposes these irregularities in the rotation of the Earth are too large. For example, an irregularity of 5 milliseconds in the length of the day represents a variation in rate of about 1 part in 10^7 and there are many applications in radio communication, surveying and radio navigation which use frequencies which must be correct to 1 part in 10^7 or 10^8. There are indeed a number of applications which require even greater precision and it is clear that if our standard of time is to keep pace with modern scientific and technical developments then we must accept the fact that the Earth is not a good standard clock.

Standard time

We have already seen that the improvement of clocks in the 17th century was followed by a change in the system of civil timekeeping from apparent solar time to mean solar time. For many years it was *local* mean solar time which was used, but in the 19th century as transport and communications improved and the use of railways spread, it became more and more desirable that large areas of the world should share the same time. To cope with this problem an international conference in 1884 agreed that the world should be divided into 24 *Standard Time Zones*. These zones were numbered in sequence, with opposite signs east and west of Greenwich and the *Standard Time* in any zone was found by adding, with the correct sign, the number of the zone to the local mean solar time at Greenwich. For many years the time at Greenwich was known as Greenwich Mean Time but more recently its origins have been hidden under the anonymous, but internationally acceptable, title of Universal Time (U.T.).

Where the difference between Standard Time and Universal Time reaches 12 hours a line was drawn on the map, the *International Date Line*, which runs across the Pacific Ocean on the opposite side of the world to Greenwich at longitude 180°. Crossing this line from east to west advances the date by one day; crossing from west to east, as Phineas Fogg forgot, reduces the date by one day. Many people find it hard to believe that this system really works; it seems so unnatural to pass from Monday to the previous Sunday without anything visible happening. As one might expect, it was not introduced without argument and in the case of Tonga, the tidy-minded lost the battle. Tonga is about 5° of longitude, (20 minutes of time) to the east of longitude 180° and ought therefore to be on the east side of the International Date Line and its Standard Time should be 12 hours *after* Greenwich. But the Tongans wanted to be ahead of everyone else; they wanted 'time to begin' in Tonga. So they chose to be on the *west* of the Date Line, putting a nasty kink in it, and they made their Standard Time, not 12 hours behind Greenwich, but 13 hours *ahead* of Greenwich. One of the supporting arguments for this decision proved compelling; the King pointed out that on the World Day of Prayer the Tongans would be the first people on Earth to say their prayers. There was another minor religious complication; the Seventh Day Adventists, who elsewhere take Saturday as their sabbath, decided to hold it on Sunday because Tonga should really be east of the Date Line.

The system of timekeeping in use in the 19th and early 20th centuries was overtaken by developments in electronics and a critical point was reached in the early 1940s when the time intervals derived from a quartz oscillator were shown to be significantly more regular in length than the intervals derived from the spinning Earth. The application of the quartz oscillator to communications, navigation and timekeeping soon brought an urgent demand for a standard of time significantly more uniform than Universal Time.

In an attempt to meet this demand astronomers put forward the suggestion that the standard unit of time, the second, should be defined in terms of Ephemeris Time. This is an astronomical system which, in effect, uses the whole solar system as a clock and is based on the orbital motions of the planets round the sun. It is completely independent of the rotation of the Earth and has the merit that it is believed to be perfectly uniform. Rather surprisingly, and as it turns out, unwisely, the international definition of the second was changed in 1956 from being simply 1/86400 of a solar day to an Ephemeris second or 1/31556925·9747 of the length of the tropical year 1900. But, although the second was standardized in this way, the system of time in general use, Universal Time, was left untouched and Ephemeris Time was kept discreetly in the background for scientific purposes.

The choice of Ephemeris Time as an international standard turned out to be both premature and impractical. It was soon overtaken by a spectacular development in timekeeping, the atomic clock. By 1967 it was generally agreed that, as a standard source of time interval, nothing could beat the atomic clock. Thus, although Ephemeris Time is believed to be as uniform as atomic time, it is much more difficult to measure. In fact it can only be established retrospectively because it takes months or years to measure it with very high precision and, although it is important astronomically, it is of very little use as a standard of time for general scientific purposes. In contrast it is relatively simple to establish the unit of time, and hence frequency, with an atomic clock which can be transported easily, or reproduced anywhere in the world. It is therefore not surprising that in 1967 the international standard unit of time, the atomic second, was defined in terms of the caesium atom. The Système Internationale (SI) second is now the duration of 9 192 631 770 periods of the caesium atom. This number was chosen so that, as far as possible, the duration of the second remained unchanged from the previous definition in terms of Ephemeris Time.

This redefinition of the second has led to a curious situation in which our present-day system of timekeeping is no longer based wholly on the sun. The time shown by our sitting-room clock, is no longer Universal Time (U.T.) but is called Universal Co-ordinated Time (U.T.C.). This new system, has two essential features; firstly the basic unit of time interval is the standard (S.I.) second which is derived from an atomic clock and is therefore uniform in length and independent of the vagaries of the Earth; secondly the actual time shown by our clocks is kept in step with U.T. and hence with mean solar time. As the rotation of the Earth slows down, the *mean solar second* is getting longer than the *atomic* (S.I.) *second*, and so to keep our clocks in step with U.T., and hence with the sun, it is necessary to correct them occasionally. The corrections are made by occasionally missing one complete second in such a way that the difference between our clocks (U.T.C.) and U.T. is never more than 0·75 seconds. How often these 'leap seconds' are made depends on how steady the rotation of the Earth proves to be; if they are needed they are made at the end of June or December and are advertised in advance.

In passing it is interesting to note that there are now two systems of time interval which are both believed to be uniform—International Atomic Time (T.A.I.) based on the caesium atom, and ephemeris time (E.T.) based on the solar system, one promoted by physicists and engineers and the other by astronomers. It will be interesting to see whether or not these two systems show signs of diverging over a period of many years. For example, if the force of gravity is really decreasing with time, as some people have suggested, then the two systems will diverge. At the present time they appear to be keeping in step, but the observations only extend over a few years. If they do change with respect to each other, it is not immediately obvious how we shall decide which of them is changing!

The extraordinary mixture of atomic and solar time (U.T.C.) which we now use is an attempt to get the best of both worlds, the worlds of physics and astronomy. On the one hand the basic unit of our system, the second, is precise, invariant, internationally standard, and therefore reliable in a wide variety of physical measurements where exact and reproducible frequencies are required. On the other hand over long intervals like a year, the time shown by our clocks keeps in step, to within 0·75 seconds, with U.T. based on the sun.

As we saw in the last chapter, the interests of society in the year and the day could only be reconciled by introducing the leap year. It is of great interest to find that, in much the same way, the interests of society in the second and the day can only be reconciled by introducing the leap second.

Our present system of timekeeping does not, of course, please everybody. As far as the man in the street is concerned there is no difference between the present system (U.T.C.) and the previous system (U.T.), it makes no significant difference if one boils an egg for $3\frac{1}{2}$ atomic minutes or $3\frac{1}{2}$ mean solar minutes. But there is some division of opinion among specialists. The need to take into account the difference between our clocks on U.T.C. and U.T., which may amount to 0·75 seconds and which varies between leap seconds, makes the system unattractive to people who are concerned with precise measurements relative to the rotating Earth. Such people include astronomers, surveyors, and some navigators. On the other hand the system suits people who are concerned with precise measurements involving time or frequency, such as physicists, electronic engineers, and those who are concerned with radio aids to navigation.

One wonders how long the present standard will last. At the present time the rate of a caesium clock can be made stable to about 1 part in 10^{13} over long periods and standard time in different countries can now be synchronized, using radio, satellites, or even travelling atomic clocks, to a few microseconds. One might have thought that this was good enough, but there are, nevertheless, even more stable clocks under development which promise ten or even a hundred times better stability than a caesium clock. If they are successful then they will surely create a demand for even higher precision in the standard of time. Where this demand will arise and how much practical use this higher stability will have, one cannot be sure. For example, the demand might come from the need to keep national standards of time in different countries more closely in step with each other or from the development of navigation. Indeed it was the needs of navigation which stimulated the great advances in precise timekeeping in the 18th century and, more recently, it has been the development of precise methods of radio navigation of aircraft, ships, and submarines which has encouraged the development of sources of precise and stable frequency. Perhaps the needs of navigation have not been satisfied yet. For example, a system of avoiding aircraft collisions has been proposed which relies entirely on the widespread use of very precise time. Every aircraft would carry a source of time and all aircraft would emit radio pulses simultaneously. Any aircraft could then tell how far away its nearest neighbours were by simply timing the pulses received from them and by displaying this information in some suitable way. For such a technique to work the aircraft clocks must be synchronized to better than 0·1 microseconds, and this would place new demands on the precision of worldwide timekeeping. Once more the system may have to change.

Our original question was 'What is time?' and we decided to ask instead, 'How is time measured?' We found that in our present day, time is based on both the atom and the Earth. The ticking of our clocks, the length of our seconds, is regulated by counting the vibrations of the caesium atom; while the actual time shown by our clocks is kept in step with rotation of the Earth relative to the sun. Curiously, we count the very small and the very large. And so the answer to our second question is that time is measured by counting a sequence of events. We count the swing of the pendulum, the vibrations of an atom, and the revolutions of the Earth, and we call the count time.

Perhaps the great philosopher and mathematician Gottfried Leibnitz was right when he said, 'Space is the abstract of all relations of co-existence; time is the abstract of all relations of sequence'. Nevertheless, I would prefer to leave our original question 'What is time?' unanswered. Let us leave it to the imagination where it belongs and not add to the load of metaphysical speculation with which so many books are burdened. Intuitively, one feels that there is something more fundamental to be said about time but, if there is, it is elusive; no doubt that is why people like Aristotle and St Augustine have written so much. But the history of science warns us not to trust to intuitive ideas about motion, and space, and time, for they are apt to be completely wrong. We have only to reflect on Newton's laws of motion (see Chapter 4) to realize how true this is, and there are many other examples nearer to our own time. For example, the quantum theory, the wave-particle nature of light and the time paradox of relativity (see Chapter 4) are all awful warnings that intuition, or if you prefer, common sense, is not a good guide to the nature of the physical world.

3 Stars and Navigation

The ability of some animals to find their way about the surface of the Earth is barely credible. Migrating birds travel thousands of miles to their own particular nests, salmon return to the streams where they were hatched after years spent in the open sea, and it looks as though the green turtle can swim across 2000 km of ocean from the east coast of Brazil to Ascension Island. There is much about these astonishing feats of navigation which is not yet understood. For example, one has only to consider the classic case of one particular bird, a Manx Shearwater, to realize that this type of bird doesn't find its way about simply by recognizing familiar landmarks. This particular Shearwater was taken by aircraft from its burrow on the island of Skokolm, Wales, to Boston and released. It returned to the same burrow, covering a distance of about 5000 km, in 13 days. Most of this journey was over water and clearly the bird did not steer by familiar landmarks; it must have had some other aid to navigation.

Experiments on migrating birds suggest that they use a variety of aids to navigation, which include the Earth's magnetic field and the direction of the sun and stars. But, as we have already seen in Chapter 2, the direction of the sun is not a simple signpost. At local noon it is either due north or south, depending on which side of the equator we are, but at any other time its direction depends upon the time of day, the date, and the latitude of the observer. Just how a bird manages to take all this into account is not known. Perhaps it knows by experience, from the season and the height of the sun, which way it should fly. Experiments in a planetarium have shown that birds can also use the stars as a guide. Apparently they can recognize patterns of stars and find their way in relation to those patterns.

We have only to look at a map of Oceania to realize that for a very long time man also used the sun and stars as a guide. The enormous expanses of ocean between the islands of the Pacific have been crossed and recrossed by sailing canoes for centuries and this has all been done without the modern paraphernalia of navigation, in fact most of it without any instruments at all. The primitive art of sailing by the sun and stars must have reached its highest development in the Pacific where an error in navigation is infinitely more serious than in the Mediterranean where one is never far from land; nevertheless, it is in the countries bordering the Mediterranean that we must look for the origins of modern navigation, a development which is inseparable from the history of astronomy.

The shape and size of the Earth

I suppose that the first step in the development of a scientific system of navigation is to make a map, and to do this one must find out the shape and size of the Earth and how to represent its spherical surface on a flat map. The answers were known to Greek astronomers centuries before the birth of Christ, but it took a long time for them to be put to any practical use. In retrospect it is hard to realize how difficult some of these early ideas were to grasp. Thus we laugh at the ancient and widespread belief that the Earth is flat, but in earlier times the suggestion that it is shaped like a ball must have seemed even more ridiculous. As Lactantius, tutor to Constantine's son, wrote in his *Divinae Institutiones* (c. A.D. 310), 'About the antipodes one can neither hear nor speak without laughter. It is asserted as something serious that we should believe that there are men who have their feet opposite ours . . . Is there,' he asks, 'anyone so foolish as to believe that . . . there is a place where the trees grow downwards and where it rains, hails and snows upwards?'.

The idea that the Earth is spherical must have been advanced many times before it was generally accepted. We know that it was taught by Pythagoras in the 6th century B.C. and, later, it is to be found in Plato's *Phaidon*. However, it probably received its first thorough treatment in the writings of Aristotle, the most influential of all the ancient writers on science, who claimed in his *De Coelo* in about 350 B.C. that the Earth must be spherical because the outline of its shadow on the moon during eclipses was circular —a single eclipse, by the way, is inconclusive; the Earth could be cylindrical! As supporting evidence Aristotle pointed out that travellers to the north reported that some stars rose higher above the horizon and some were lost below the southern horizon. Some of Aristotle's other arguments are attractive but not so sound. For example, he also argues that the Earth must be round because elephants had been seen to the east in India and also to the west in Morocco and therefore these two places must be joined.

I suppose the first striking confirmation that the Earth is round came when Magellan sailed west from Seville in 1519 and his ship Vittoria returned, without Magellan, from the east in 1522. Nevertheless, the answer to our first problem is that the shape of the Earth was established by sound astronomical arguments over 2000 years ago.

But what about the size of the Earth? If we turn again to the Greeks then, as every book on astronomy tells us, the size of the Earth was measured by the astronomer, Eratosthenes of Cyrene, in about 240 B.C. He did this by measuring the difference between the angle of elevation of the sun above the horizon, the so-called *altitude* of the sun, as seen at noon on the same day in Syene (Aswan) and Alexandria. Syene is on the Nile about 800 km south of Alexandria, almost on the tropic of Cancer; at the summer solstice the sun is therefore very close to the zenith and it is said that its image was reflected in the water at the bottom of a deep well. On the same day in Alexandria Eratosthenes found that the sun is $7\frac{1}{2}°$ from the zenith at noon. The difference in latitude between the two places is therefore $7\frac{1}{2}°$ and, from measurements of the distance between them, he was able to work out the distance on the Earth's surface corresponding to $1°$ in latitude. Multiplying this result by 360, he found the Earth's circumference to be 252 000 stades. There is some uncertainty about the length of a stade which was not firmly standardized throughout Greece; however, on the assumption that there were 10 stades to a Roman mile (1488 m) this value for the circumference of the Earth was only about 7 per cent less than our modern value.

Thus, both the size and shape of the Earth were known to astronomers 2000 years ago and it was not until the seventeenth century that, to meet the demands of navigation, more precise observations were made.

The first scientific map

If we look at a collection of ancient maps of the world the first which looks really modern was published by Ptolemy during the second century A.D. Ptolemy was one of the last great figures in Greek astronomy and is best known as the author of the *Almagest* which, for over a thousand years, was regarded as the leading text on astronomy. Nowadays, we are not so sure about the *Almagest*; there is a good deal of evidence that some of the 'original' astronomical data in it was not only borrowed but fudged! It seems that Ptolemy may have 'adjusted' observations to fit his theories. However that may be—a fascinating problem in the history of science— there is little doubt that Ptolemy's map of the world was a real step forward because it was based firmly on astronomical science; it owes a lot, like the *Almagest*, to Hipparchos, who lived in the second century B.C. In this map the Earth is assumed to be spherical and the surface is marked out in the modern way using latitude and longitude and a conical projection. Unfortunately, the scale of distance was taken, not from Eratosthenes, but from a later astronomer, Poseidonios (135–51 B.C.), who is reported to have measured the length of a degree of latitude between Alexandria and Rhodes. Again there is some confusion about the scale. One author reports that Poseidonios found the circumference of the Earth to be 240 000 stades, in reasonable agreement with Eratosthenes, but another reports 180 000 stades. It has been suggested that this discrepancy can be explained by differences between the length of the stade used at different times and places. Nevertheless, Ptolemy took the value as 180 000 stades.

It is surprising to find that Ptolemy's map of the world, published in the 2nd century A.D., was still important in the 16th century. This was, of course, true of many other features of rediscovered Greek science. One must also remember that in the centuries which separate Ptolemy from

Map of the world according to Ptolemy. This version was drawn in 1730.

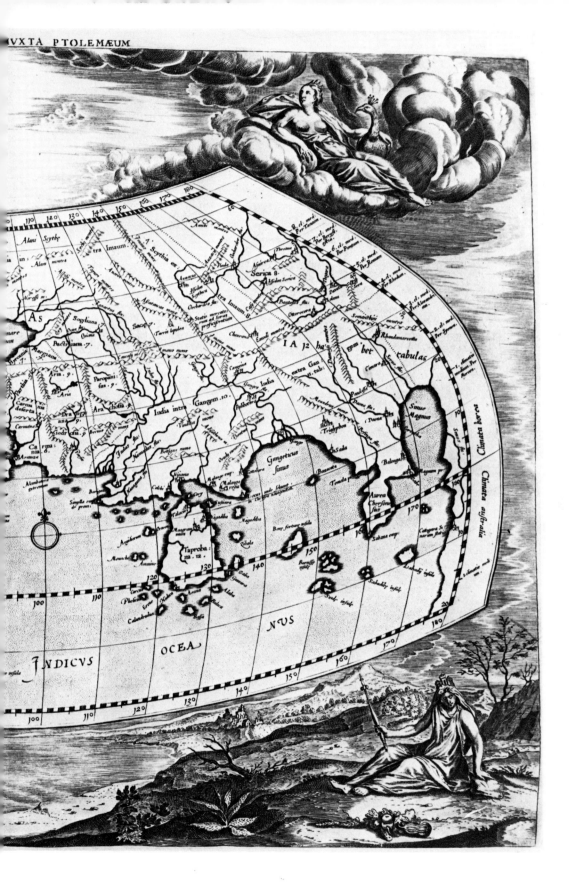

Columbus most new maps were limited to areas of interest to traders and were based on measurements of distance and bearing made by dead reckoning. As we shall see, some use was made of astronomical measures of latitude in the mid-15th century but it was not until the end of that century that the great voyages of exploration made realistic maps of the whole world necessary. Before then, mediaeval maps of the world were largely of academic interest and some of them were made by monks who were inclined to seek guidance from the Bible and not from astronomers. Unfortunately for the progress of cartography, the Bible is not a reliable guide; depending on where we look, it tells us that the Earth has four corners, that it is circular with Jerusalem at the centre, and so on, but it does not make the vital point that the Earth is a sphere. Most mediaeval maps of the world therefore look ridiculous to a modern eye; they are rectangular or circular with all the land in the middle surrounded by water.

In comparison, Ptolemy's map of the world is strikingly scientific and, when it appeared in Europe with the translation of his works from Greek into Latin in 1409, it was welcomed by an age in which an interest in mathematics, geometry and astronomy was reviving. The development of modern, scientific, navigation dates from the time when navigators and map makers started to think of their position in terms of latitude and longitude rather than distance and bearing from prominent landmarks. Ptolemy's map was an important influence in bringing about that change.

The early measurements of latitude

The latitude of a place on the Earth, as we saw in chapter 1, is its angular position from the poles and is measured as the number of degrees north (N) or south (S) of the equator; the longitude is its angular position relative to the Earth's axis of rotation and is measued east (E) and west (W) of some particular meridian of longitude, nowadays the meridian of Greenwich. If we project the surface of the Earth on to the inside of a hollow sphere surrounding the Earth then on this sphere latitude is called *declination* and longitude is called *right ascension*. These are just two ancient names used for the latitude and longitude of stars on the celestial sphere and there is nothing mysterious about them.

This simple correspondence between positions on the Earth and in the sky means that if we look vertically upwards at night we see stars which have a declination equal to our latitude; for example, if we are in London at latitude $51\frac{1}{2}°$N then the stars which pass vertically overhead all have the same declination $51\frac{1}{2}°$N. Thus, in principle, it is elementary to find one's latitude; look vertically upwards, note which stars pass overhead, and then find the declination of those stars from a map of the sky. Roughly speaking, that is what primitive navigators were doing when they were lying on the

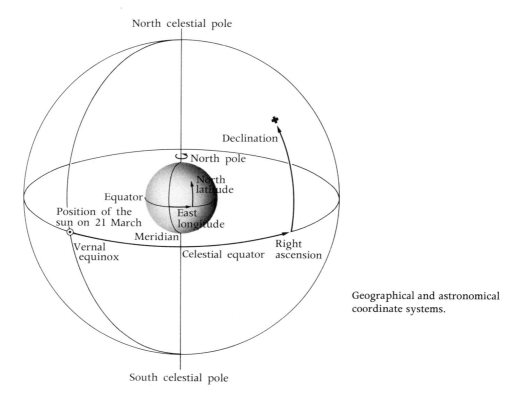

North celestial pole

Declination

North pole

North latitude

Equator

East longitude

Position of the sun on 21 March

Meridian

Vernal equinox

Celestial equator

Right ascension

South celestial pole

Geographical and astronomical coordinate systems.

deck of the boat squinting up the mast.

The earliest methods of finding latitude were based on the pole star and the sun. If only the pole star were precisely at the north celestial pole then measuring latitude would be child's play; all we would need to do would be to measure its angle of elevation above the horizon—its altitude—and that would be the latitude. The difficulty is, of course, that the pole star is not exactly at the true pole and, even if it were, it wouldn't stay there for long because the Earth's axis slowly precesses round in a circle (see Chapter 1). Nowadays the pole star is less than 1° from the true pole but in the late 15th century, when systematic measurements of latitude really began, it was about $3\frac{1}{2}$° from the pole. An error of $3\frac{1}{2}$° in latitude corresponds to an error of roughly 400 km in position on the Earth and so, even by the relatively crude standards of 15th century navigation, it was necessary to take into account the fact that the pole star is not exactly at the true pole.

In the course of 24 hours the pole star circles the true pole and so the difference between its altitude and that of the true pole depends upon the local sidereal time (see Chapter 2). However, in any 24 hours there are two times, 12 hours apart, when the pole star is at exactly the same altitude as the true pole and the simplest method of finding latitude is to measure its altitude at one or other of those times. To do this one must recognize what the sky looks like at those times and, with a little practice, one can learn to use the position of the Little Bear or the Great Bear. To make it easier, a simple set of rules was developed in the latter half of the 15th century known as 'The regiment of the north star'. The earliest surviving written directions are to be found in the *Regimento do Astrolabio e do Quadrante* published in 1509. A more sophisticated method, used in the 16th century, was to use a single instrument with a pointer and scale which gave the position of the true pole directly. The navigator aligned this instrument on the pole star by sighting through a hole in the centre of the scale and then rotated the pointer to align it with the stars in Ursa Minor. The pointer then showed directly on the scale the angular distance and direction of the pole from the pole star.

Latitude can also be found by measuring the altitude of the sun at noon; if you know the declination of the sun, a simple sum then yields your latitude. However, as with the pole star, there are complications due to the fact that the sun is always moving with respect to the background stars and so its position on the celestial sphere (declination and right ascension) is also changing. At the equinoxes, in March and September, its declination varies by about $\frac{1}{3}°$ every day and, since an error of $\frac{1}{3}°$ in finding latitude corresponds to an error in position of about 20 nautical miles (37 km), it is obviously important to know the declination of the sun on the actual day on which you want to find your position. This requires tables of the sun's position which were not available in a convenient form for navigators until the end of the 15th century.

Arabic navigators in the Indian Ocean developed quite advanced techniques of finding their way about. After all, they had been doing much the same thing in their deserts for centuries. However, the systematic use of measurements of latitude as an aid to the navigation of ships dates from the Portuguese voyages of exploration in the 15th century. An important driving force behind these voyages was Prince Henry the Navigator, who as Grand Master of the Order of Christ, devoted himself, and a substantial fraction of the wealth of the Order, to the maritime exploration of Africa. No doubt this was one of those enterprises in which the religious motives, seeking new converts among the heathen, were happily combined with a love of adventure and a search for wealth. By rounding the tip of Africa, if there proved to be a tip, he hoped to outflank the Moslem infidels of the Near East and even link up with the legendary Christian kingdom of Prester John. At the same time it would clearly be commercially profitable to find a sea route to the Indies and thereby cut out the Arabian middle-man in the spice trade.

These voyages, whatever their motives, mark the start of the systematic development of navigation. As Prince Henry's ships worked their way slowly down the west coast of Africa they ventured farther and farther from the coastlines which they knew, and it became increasingly important to develop a more reliable means of fixing position than elementary dead-

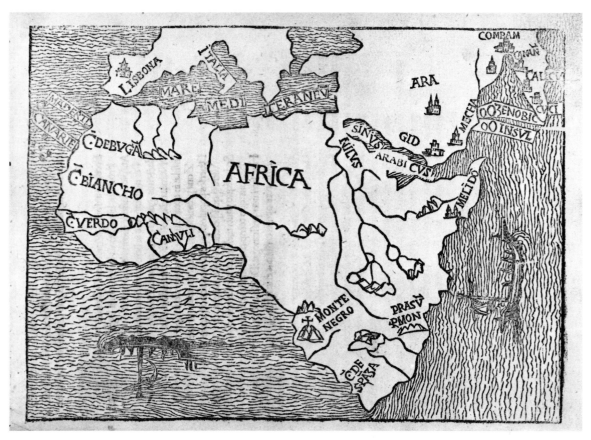

A Portuguese map of Africa made in 1508, ten years after Vasco da Gama
had sailed round the Cape to India.

reckoning based on a compass and an estimate of the distance travelled—a
system which was at the mercy of ocean currents. Not only did the
Portuguese need to fix the positions of their ships, but they wanted to put
the places which they discovered on the map, and so, as there was no
practicable method of measuring longitude and as their voyages were
mainly north and south, they concentrated on the measurement of latitude.
In fact, they did not round the tip of Africa in Prince Henry's lifetime (he
died in 1460), but by 1474 they had compiled a list of latitudes, accurate to
about $\frac{1}{2}°$, reaching to the equator.

These early measurements of latitude were based, almost entirely, on
observations of the pole star. However, the crossing of the equator in 1480
raised a new and urgent problem. South of the equator the pole star sank
below the northern horizon and, until navigators learned to use other stars,
they had to rely on the sun to find their latitude. Up to that time
observations of the altitude of the sun at noon had been used largely to

51

measure small changes in latitude over voyages of a few days. But, as we have just seen, to make absolute measurements of latitude it was necessary to know the sun's position and this demanded new tables giving the sun's declination to the nearest $\frac{1}{2}°$ for every day of the year. Such tables did not exist in a suitable form and to meet the demand King John II of Portugal appointed a special commission in 1484 which in due course sponsored a set of simplified tables of the sun's daily position based on the work of the astronomer Zacuto of Salamanca. It is interesting to note in passing that about 300 years later another special commission in Great Britain, the Commissioners of Longitude, sponsored a set of tables of the moon's position to meet the demands of finding, not latitude, but longitude. The tables of the sun's position were published in the *Regimento do Astrolabio e do Quadrante*; they made it practicable to find latitude from any place on Earth from which the sun could be seen.

It was not long, however, before the Portuguese studied the southern sky and learned to find their latitude by observing the Southern Cross. In 1505 John de Lisboa published some simple instructions: 'Hold a plumb line in front of you and when the ''head and foot'' of the Southern Cross are in line with it, then the foot (Acrux, α Crucis) is 30° above the true south celestial pole'. This is no longer true, due to precession of the Earth's axis, but it was true in the 16th century and afforded a simple method of finding latitude in the southern hemisphere.

The need for longitude

Most of the long voyages in the 15th century, at least the ones which led to the systematic development of navigation, were along the coast of Africa and could be made without worrying about longitude. Because the west coast of Africa runs mainly north and south, a ship could sail south, safely clear of the coast, until it reached the latitude of its destination, and then turn east and make a landfall. This is more or less what Bartholomew Diaz was doing in 1487 when he set sail from a place on the west coast of Africa, ran south for 13 days before a strong wind, and turned east to make a landfall. He found no land until he eventually turned north; he had rounded the Cape of Good Hope and found a sea route to the East.

This voyage of Diaz was soon followed by the discovery of the West Indies by Columbus in 1492 and the voyage of Vasco da Gama to India in 1497. A scramble for new territories and trading rights started among the major maritime powers, and the improvement of navigation became a matter of national concern. Better charts were drawn, better navigational instruments were made, almanacs were printed and schools were founded to train navigators; the help of astronomers was enlisted to calculate tables of tides, and positions of the sun, moon, planets, and some bright stars. A problem which had not been solved satisfactorily was how to measure longitude. The Indies to the east and the New World to the west involved voyages with long stretches of east–west sailing out of sight of land, in fact on his first voyage to the West Indies in 1492 Columbus was out of sight of land for 33 days. Under these conditions the old techniques of coastal sailing were no longer adequate, and without some means of finding longitude, other than dead reckoning, the explorer was never sure how far to the east or west he had sailed, nor could he fix with certainty the position of any new lands which he discovered.

The difficulties of not being able to measure longitude are illustrated by the controversies about the relative spheres of influence of Spain and Portugal. Following the discovery of the West Indies by Columbus, the Pope, with staggering self-confidence, divided between these two countries 'all the firm lands and islands found or to be found, discovered or to be discovered toward the west and south'. In the absence of any means of determining longitude he specified the position of the dividing line by its distance in leagues west of the Azores. The final agreement about this line was reached in 1494 in the Treaty of Tordesillas which laid down a north–south line 370 leagues west of Cape Verde, in modern terms somewhere between 40° and 50° west of Greenwich. Everything to the west of that line lay in the sphere of Spain, everything to the east was assigned to Portugal. This arrangement settled the arguments for some time, but not for long because, as one might expect, trouble arose over where the dividing line was on the other side of the world. The most active argument was about the Spice Islands; were they in the region assigned to Portugal or Spain? As the longitude of the Spice Islands is roughly 125°E of Greenwich they were somewhere between 165° and 175° of longitude from the Pope's line, uncomfortably close to half way round the world. To settle this question purely by measuring distance, and not longitude, demanded a reasonably accurate knowledge of the circumference of the Earth, about which there was considerable disagreement.

From the point of view of an astronomer the interesting feature of the controversy which followed is that it reflected the classical discrepancy between the circumference of the Earth found by Eratosthenes in 240 B.C. and the value adopted by Ptolemy from the work of Poseidonios. The Spanish took their value from Ptolemy and argued that the distance of the Spice Islands, measured in leagues east of the Pope's line was greater than half the circumference of the Earth and therefore they were definitely not in the Portuguese sphere of influence. The Portuguese took their value from Eratosthenes and argued that the islands lay within their sphere, and indeed they were right. The dispute was not finally settled on scientific grounds but by some hard trading in which it was agreed that the Portuguese should control the Moluccas but the Spanish should control the Philippines.

In the course of the 16th century, trade with these newly discovered lands grew to such proportions that it was a significant factor in the economy of Europe. As the value of the cargoes increased so did the importance of safe navigation and hence of finding longitude. Many handsome prizes were offered for a solution to this problem, one of the first being the prize of 100 crowns offered by Philip III of Spain in 1598. As we shall see, however, although the problem of finding longitude was recognized and great prizes were offered, no solution was forthcoming until the middle of the 18th century.

The problem of finding longitude

As we have already seen, it is simple to find latitude; all you have to do is to note which stars pass directly overhead and then look up their declination on a map of the sky. At first sight it looks as though we ought to be able to find longitude in much the same way; we can't, because the Earth is spinning. The essential difference is, of course, that latitude is measured with respect to the equator and poles of the Earth which remain stationary with respect to the stars; longitude, on the other hand is measured from some arbitrary north–south line, nowadays through Greenwich, and this line is not fixed with respect to the stars but rotates with the Earth. This means that to measure longitude by the stars we are concerned essentially with the rotation of the Earth or, in other words, with the measurement of time. Thus, the simplest way of finding the longitude of a place is to measure the time interval between a star crossing the meridian (north–south line) at that place and at Greenwich. For example, if a star crosses the meridian at Rome one hour before the meridian at Greenwich, then the longitude of Rome is 15° east of Greenwich because in 24 hours the Earth rotates through 360° and therefore in one hour rotates through 15°. To make such a measurement a navigator would need a clock giving the time at Greenwich and also an almanac from which he could work out the time at which that particular star crossed the meridian at Greenwich. Although this idea was put forward many times in history, for example in 1522 by Gemma Frisius, a Professor at Louvain, it was completely impracticable until the 18th century because there was no timepiece accurate enough to measure longitude on land, let alone at sea. To find one's longitude to the nearest $\frac{1}{2}°$ (30 nautical miles at the equator), one must know the time at Greenwich to within 2 minutes and, if the voyage takes a month, the clock must keep time to about 4 seconds a day. In Tudor times a clock could be expected to keep time to about 15 minutes a day and it was not until the early part of the 18th century that the very best pendulum clocks, which certainly could not be taken to sea, achieved an accuracy of 4 seconds a day.

Thus, before the 18th century, the only possible techniques for measuring longitude at sea were those which had been used on land to build up the map of the world. The oldest of these methods, which was known to the Greeks, was to observe the local times at which an eclipse was seen at two places. The difference between these times gave the difference in longitude. Although it is simple and reliable and was used extensively on land, the use of eclipses was obviously of little use to ships; eclipses are rare events, and, to make matters worse, only eclipses of the moon can be used because eclipses of the sun are seen from only a small fraction of the Earth's surface.

There is, however, another way in which the motion of the moon can be used. Longitude can be found by measuring the angular separation, also called angular distance, between the moon and a star or between the moon and the sun; the local time of the measurement is noted and then, with the aid of an almanac, the time corresponding to that particular angular distance is computed for some place of known longitude, e.g. Greenwich. The difference between these two times gives the required difference of longitude. This technique, later to prove successful, is called the method of lunar distances or *lunars* and, like the use of a transportable clock, it was also put forward long before it could be implemented. The well-known German astronomer Regiomontanus suggested it in 1475, but until the 18th century no one could exploit it for two good reasons. First, there was no instrument with which the angular distance between the moon and a star could be measured with sufficient accuracy from the deck of a moving ship; secondly, astronomers could not predict the relative positions of the moon and the star with the necessary precision.

A third way of finding longitude is to watch the satellites of the planet Jupiter. Thus if we watch one of the satellites, say Io, as it revolves around Jupiter, we can use it as a clock which everyone, or at least everyone on the same side of the Earth, can see. If we note the local time at which Io appears or disappears behind Jupiter and compare this with the time at which it appears to do so at Greenwich, then the difference between these two times gives us our longitude. If we are not in a hurry to know the result, then this procedure can be carried out by two observers, one at each place. But in the case of a ship at sea it would be necessary to carry a table giving the predicted Greenwich times at which Io appears and disappears on that particular date.

Although Galileo thought of this ingenious idea and offered it to the Dutch and Spanish governments in the hope of winning one of the prizes for finding longitude, it was rejected. One must remember that it involved the use of an unwieldy telescope on the deck of a ship, that Jupiter was not always in a suitable position to be observed, and that individual observations of the appearance and disappearance of the satellites are not easy to time accurately; the satellites do not disappear abruptly behind the

A detail from Galileo's notebook for 1610 showing his records of the positions of Jupiter and its satellites.

planet but fade away, and the exact time of disappearance depends on the viewing conditions and on the power and quality of the telescope. Nevertheless, under carefully controlled conditions and with skilled observers, one at each station, this method gave quite accurate results on land, but it was definitely not suitable for use at sea.

Despite these difficulties some of the early navigators did try to measure longitude by astronomical means. For example, Columbus tried, without much success, to find his longitude by observing eclipses. His successor, Amerigo Vespucci, made quite a reasonable estimate of his longitude in 1499 while off the coast of South America by observing a conjunction of the moon and Mars and finding the local time of this event in Ferrara in an almanac. As a later and more advanced example, in 1612 the explorer William Baffin made the first recorded measurement of longitude by a British navigator. Working on shore in Greenland he measured the elevation of the sun at the instant that the moon crossed the meridian and, with the help of an almanac, calculated his longitude; this was not an easy thing to do and it is not surprising to find that Baffin was wrong by about 8°. There are many other examples of attempts to measure longitude in the 16th and 17th centuries but none of the results was reliable. The plain fact is that, although progress was made in improving the accuracy of measuring longitude on land, no practicable method of finding longitude at sea was developed until the latter half of the 18th century. Most of the great voyages of exploration were made and most of the world was discovered without any method of measuring longitude.

Progress in navigation and astronomy in the 16th and 17th centuries

Apart from this question of longitude, to which no one had the answer, important advances in navigation were made in the 16th and 17th centuries and they went hand in hand with advances in astronomy. If one looks in a museum the first thing one notices is how greatly the standard of workmanship of both navigational and astronomical instruments improved, but there were also improvements in function; a significant feature of the time is that serious efforts were made to make astronomical instruments more suitable for use at sea. The three principal instruments used by early navigators to make astronomical observations were the cross-staff, quadrant, and astrolabe. All of them, in one way or another, were unsuitable for use at sea and had to be modified. For navigation at sea, the cross-staff was shortened and fitted with three cross-pieces for different angular ranges. To measure the elevation of a star the navigator viewed one of the cross-pieces from the end of the staff and slid it along the staff until its apparent angular size was equal to the angle between the star and the

Taking a sight of the pole star with a cross-staff, 1545.

Mariner's astrolabe, *c*.1585, measured latitude at sea.

horizon. The altitude of the star was then read off a scale of the staff. I have never tried to use a cross-staff but it looks difficult enough to get an accurate result on land, let alone on the moving deck of a ship and it doesn't surprise me to read that the early Portuguese explorers went ashore, whenever they could, to measure their latitude by the pole star. It must have been a most unpleasant, positively injurious, instrument to use on the sun. Another, not so obvious, objection to the cross-staff is that, in its simple form, it could only be used to measure a limited range of angles. The lower limit, set by the size of the whole thing, was usually about 10° and the upper limit, set by what one can do with the eye, was about 60°. In practice this meant that a cross-staff could not be used to measure lattitude near the equator, because the sun was too high in the sky and the pole star was too low. Likewise, in high northern latitudes, where so many of the early British explorations were carried out, the sun was too low in the sky and this was important because the sky was too bright on summer nights to use the stars.

It was in order to get over these difficulties that Captain John Davis invented the backstaff in about 1595. In its simplest form this consisted of a long graduated staff on the end of which was mounted a plate with a

An astronomer using an astrolabe to measure the altitude of the stars, helped by a mathematician and ▷ a clerk. A painting in a 13th-century manuscript.

58

A backstaff, invented by Captain John Davis in about 1595. This one was made around 1740.

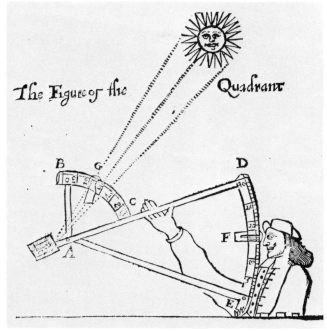

Measuring the altitude of the sun using a backstaff.

horizontal slit in it. A semicircular cross-piece slid on the staff. The observer stood with his back to the sun and looked along the staff holding it so that he could see the horizon through the horizontal slit; he then moved the sliding arm until the end of its shadow fell on the slit. Compared with the classical cross-staff this was a great advance; the observer was not blinded by the sun and he was only required to look in one direction, horizontally, to make a measurement. In a slightly more complicated form the backstaff could be used to measure altitudes up to 90°, thereby solving the problem of measuring latitudes by the sun near the equator. It remained in popular use for nearly 200 years.

There were many other improvements in navigational instruments which included more precise and rugged versions of the astrolabe and the quadrant. One particularly notable innovation was the introduction of the telescope early in the 17th century. Admittedly its principal value at that time was probably in naval warfare, where its impact has been compared with the introduction of radar in recent times, but it was also of great and lasting value to the safe navigation of ships when close to land.

These improvements were valuable to the navigator because they increased the accuracy of the data on which he worked; but of even greater value were the advances in the way in which he handled these data. Up to the beginning of the 16th century mathematics was not widely taught in the schools of Great Britain and some navigators even used Roman numerals. You have only to try a few simple sums, particularly multiplication, using Roman numerals to realize how cumbersome they are. However, in the latter part of the 16th century there was rapid progress in the application of mathematics to a variety of topics, including navigation; no doubt this was accelerated by the appearance in 1567 of the first translation of the works of Euclid on geometry into English.

A particularly significant advance was made in 1614 when John Napier invented logarithms. Their importance to navigation was at once realized by Henry Briggs, the Professor of Astronomy at Gresham's College in London, who put them into a form useful to navigators in 1616. The use of these logarithms made it possible to do quite complicated calculations both quickly and reliably, and navigators were soon solving spherical triangles and working out great circle courses. The use of logarithms made it possible to do navigational calculations without making crude approximations to simplify the arithmetic; this was an essential preliminary to the great increase in the precision of navigation which took place in the 18th century.

The great steps forward made by astronomy at this time were another essential preliminary (see Chapter 5). The first was made by Copernicus in 1543; he showed us that the Earth and the other planets revolve around the sun and that the Earth spins on its own axis. The second was made by Kepler between 1605 and 1619; he discovered the mathematical laws which govern the motion of a planet around the sun. The third was made by Newton in 1687; he showed mathematically how the laws of planetary motion, Kepler's laws, can be explained by one simple universal law of gravitation. The importance to navigation of these three discoveries lies in the fact that they mark the beginning of that mathematical, quantitative, dynamical understanding of the solar system which, in later years, enabled astronomers to meet the demands of navigation for very precise astronomical data.

Striking progress was also made in observational astronomy, notably by

Tycho Brahe. Tycho Brahe was the greatest observer of the pre-telescopic age; in the period 1576–97 he made the most precise and systematic observations of the sun, moon, planets, and stars which had ever been made. Using very large instruments of high quality and a very careful observing procedure he achieved a positional accuracy of about 1 minute of arc. His measurements are of great historical importance, not only because they were used for so many years, but because they were sufficiently detailed and precise to make it possible for Kepler to establish the laws of planetary motion. However, the most revolutionary advance in this period was the application of the telescope to astronomy by Galileo in about 1609. The work of Copernicus, Kepler, and Newton marks the birth of modern mathematical astronomy and the application of the telescope by Galileo marks the birth of modern observational astronomy.

The offer of a prize for longitude

I suppose that in the 16th and 17th centuries the fact that one could not fix the position of a ship with certainty when out of sight of land was accepted by most mariners as one of the inevitable dangers, perhaps only a minor danger, of life at sea. There were many other dangers such as scurvy or being attacked by the pirates who, in the early 17th century reached plague proportions in European waters. Nevertheless, everyone who was concerned with the efficiency of the navy or with the profitability and safety of merchantmen knew that there was a real need to improve navigation; handsome prizes for a method of finding longitude were offered by the governments of France, Holland, Spain, Venice, and finally Great Britain.

There was no shortage of ideas about how longitude might be measured; as we have already noted, the two principal techniques, lunar distances and the use of a transportable clock, were both put forward at an early date. The problem was to implement these ideas in a practical form and this required, not simply bright ideas nor the incentive of a prize, but a milieu in which navigators, instrument makers, and astronomers could work together to solve the problem. London provided all that was needed in the latter half of the 17th century and it was largely because of this that the problem of finding longitude was solved in Great Britain.

Important steps towards creating these favourable conditions were the foundation of the Royal Society in 1662 and of the Royal Observatory at Greenwich in 1675; to quote the warrant from Charles II, the Observatory was founded 'in order to the finding out of the longitude of places for perfecting navigation and astronomy'. This latter decision was taken many years before the British Government offered a prize for finding longitude, and was in response to the reports of a committee appointed by the King to investigate a method of finding longitude proposed by the Sieur de St Pierre, a member of the Court. Although the actual proposal came to nothing, it was the work of the investigating committee and especially of John Flamsteed, who later became the first Astronomer Royal, which focused the attention of important people, including the King, on the fact that the basic astronomical data about the position of the sun, moon, planets, and stars were not accurate enough for the purpose of navigation.

Another important step towards creating a favourable climate was taken when the interest of the public and Parliament was aroused by a major naval disaster attributable to poor navigation. In 1707, a British fleet returning from Gibraltar ran on to the rocks off the Scilly Islands and 2000 seamen were lost. The flagship sank with 800 men, among them the Vice-Admiral commanding the fleet, Sir Cloudesley Shovel. An unpleasant reminder that, in those days, the peril of bad navigation was just one among many, is to be found in an account of this wreck which says that Sir Cloudesley probably reached the shore alive and was murdered by a thief who stole his rings. A closer look at the history of this particular naval disaster suggests that it would probably not have been prevented by any of the means for measuring longitude which were evolved later in the 18th century; but it did stir up support for improving navigation and in 1714 the British Parliament offered a 'publick reward for such person or persons as shall discover the longitude at sea'. The prize offered was 10 000 pounds sterling for a method of finding longitude to within $1°$ or 60 nautical miles, 15 000 pounds sterling for $\frac{2}{3}°$ or 40 nautical miles and 20 000 pounds for $\frac{1}{2}°$ or 30 nautical miles. In those days this prize was a very large sum of money. One would have thought that a smaller sum would have acted as a sufficient incentive and would have proved less controversial to administer.

Why did it take so long to win the prize? The only method of finding longitude which looked promising at the time was the second of the two methods we have briefly reviewed, the method of lunars. As we noted, this method involves measuring the angular distance between the moon and a star and computing the corresponding local time at a place of known longitude, say Greenwich. But consider the precision with which this must be done. The Earth makes one revolution of $360°$ of longitude in 24 hours and therefore turns through an angle of $\frac{1}{2}°$ in 2 minutes. In 2 minutes the moon moves relative to the background stars by 1 minute of arc. If, therefore, we seek to find our longitude to the nearest $\frac{1}{2}°$, we must measure the difference between our local time and Greenwich time with an accuracy of 2 minutes which involves measuring the angular distance of the moon from the star to the nearest 1 minute of arc. Furthermore, to find the corresponding time at Greenwich we must also be able to predict the relative positions of the moon and the star to an accuracy of 1 minute of arc.

Hadley's quadrant

When the prize for longitude was offered in 1714 there was no instrument capable of measuring the altitude of a star or the angular distance between the moon and a star to an accuracy of 1 minute of arc from the deck of a ship. Most measurements made from ships were, at the least, ten times worse than that. However, it was not long before this particular problem was solved by the invention of the reflecting quadrant. In Great Britain this invention is attributed to John Hadley, a country gentleman and Fellow of the Royal Society, who put forward the design in 1731. It is certain that the same idea had occurred to Newton, but he had put it on one side. As is so often the case, it was also invented independently and almost simultaneously by Thomas Godfrey of Philadelphia. A simple outline of the instrument is shown in the diagram below. A rotatable index mirror is carried by a moving index arm and reflects light from the sun, moon, or star on to a fixed horizon glass. The upper half of this horizon glass is clear and the lower half is silvered to form a mirror. To measure altitude the navigator holds the instrument upright and moves the index arm until the image of the sun, moon, or star coincides with the horizon as seen directly through the clear half of the horizon glass. The altitude is then given by the position of the index arm which is read off a scale. The angular distance between the moon and a star can be found in much the same way by tilting the instrument into the appropriate plane. For observing the sun there are dark screens which can be interposed between the index mirror and the horizon glass.

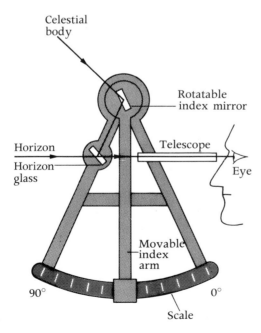

Principle of Hadley's quadrant.

This beautifully simple and efficient instrument was tested at sea in 1732 and the results showed that it was possible to measure angles with an accuracy of 1 minute of arc from the deck of a ship. This was a great advance on the performance of the Davis backstaff; it substantially improved the measurement of latitude and local time and, as we shall see, made the measurement of longitude by the method of lunars possible. Hadley's quadrant is the direct ancestor of the modern marine sextant and it is ironical that such an essential invention should not have shared in the prize for finding longitude.

The invention of Hadley's quadrant put the problem of finding longitude at sea, by the method of lunars, squarely in the hands of astronomers. To start with there are three corrections which have to be made to observations with a quadrant of the altitude of the sun or a star before the full use can be made of its accuracy. These corrections are quite minor but they do illustrate how, in order to solve some of the practical details of navigation, the answers to some of the large and apparently impractical questions of astronomy must be known. The first correction which has to be made is to allow for the height of the navigator's eye above sea level. This is a matter of simple geometry but it does involve the answer to one of astronomy's major questions, how large is the Earth? Taking, for example, the case where the navigator's eye is 12 m above the sea, the horizon seen from that height appears to be about $6\frac{1}{2}$ minutes of arc below its direction seen from sea level; thus the apparent altitude of a star is greater by $6\frac{1}{2}$ minutes of arc than its altitude seen from sea level. To compute this correction to an accuracy of 1 minute of arc we need to know the size of the Earth with an error not exceeding 8 per cent. Up to about the middle of the 17th century various values for the size of the Earth were current, some of which were in error by as much as 20 per cent. For example, it was commonly assumed by British sailors that the length of one degree of latitude or longitude at the equator was 5000 ft instead of the correct value of 6080 ft. However, by the 18th century precise and reliable values for the size of the Earth were in use. For example, the indefatigable Richard Norwood measured the distance between the Tower of London and the centre of York with a chain in 1633; he then measured the difference in latitude using an enormous sextant nearly 2 m in radius and found the length of one degree, and hence the size of the Earth, with an accuracy of about 1 per cent. Using Norwood's value for the size of the Earth, or later measurements commissioned by the French Academy, it was possible, long before the introduction of Hadley's quadrant, to compute the correction for the height of the navigator's eye with greater precision than was necessary.

The second correction is for refraction and was more difficult to get right. The path of the light from a star to an observer on the Earth is not straight but is bent downwards, refracted by the air in such a way that its apparent altitude is increased. The amount of bending varies with altitude, being greatest near the horizon and zero in the zenith. At 10° above the horizon the apparent altitude is increased by about 5 minutes of arc and at 45° by only 1 minute of arc. The existence of this effect has troubled astronomers at least since the time of Ptolemy (A.D. 150) and there is a long history of attempts to correct for it. To cut a long story short, the table of corrections produced in 1720 by James Bradley, the third Astronomer Royal, was more than adequate for correcting observations with Hadley's quadrant at sea.

The third correction, for parallax, takes care of the well-known effect that, as we move, nearby objects change their apparent direction relative to things which are farther away. This effect is called *parallax* and has to be taken into account when observing the position of the sun, moon, and planets, which are much closer to us than the stars. For example, the apparent position of the moon among the stars changes, due to parallax alone, by nearly 1° depending on whether you are looking at it when it is high in the sky or low down on the horizon. If, therefore, we wish to measure the angular distance of the moon from a star with an accuracy of 1 minute of arc, we must know the ratio of the distance of the moon to the radius of the Earth with an accuracy of roughly 1 part in 60 or, say, 2 per cent. This ratio was known approximately to Hellenistic astronomers, but it has only been known with the necessary accuracy since the Renaissance. Thus, by the time Hadley's quadrant was introduced in 1731 enough was known about the effects of the observer's height above sea level, refraction, and parallax to enable astronomers to calculate these corrections with the necessary precision.

The motion of the moon

This brings us to the major astronomical problem of finding longitude by the method of lunars, in fact to the major astronomical problem of the time; how to predict the precise position of the moon. The moon has been studied since the beginnings of civilization largely for the purposes of the calendar, but, as Charles II discovered, not enough was known in 1675 to predict its position with the accuracy required, about 1 minute of arc, to enable ships to find their longitude at sea; contemporary tables of the moon's position had errors of 10 minutes of arc or more. It has been known since antiquity that the moon's apparent motion in the sky is not as simple as it looks, and it is surprising to find that all the major irregularities in this motion were discovered by the naked eye before the telescope was invented. However, although these irregularities were recorded well enough for the purposes of the calendar, there was no sound basis for understanding them until Newton published his work on gravitation in 1687. As everyone knows, it was Newton who showed that the motion of the moon around the Earth can be explained by the gravitational attraction between them; however, he went on to show that the major irregularities in this motion can be explained by the gravitational attraction of the sun. Newton did not succeed in predicting the moon's motion with sufficient accuracy for navigation but he blazed a trail for the many talented men who followed him and completed the task. You have only to note that it took about 200 years to perfect the theory which is in use today and that it engaged the attention of mathematicians like Euler, Lagrange, and Laplace to realize that the problem was not trivial. In fact, if you look into it closely, it is obvious why only the vaguest outlines of the moon's motion are given in textbooks of astronomy and why Newton complained that it gave him a headache and kept him awake at night. Perhaps we should also note that it was not primarily the urgency of finding longitude which attracted such distinguished mathematicians to this awkward problem in the 18th century; it was the scientific importance of finding out whether or not Newton's theory of gravitation really did explain the motions of heavenly bodies; they wanted to know if Newton had, in fact, discovered the secret of the celestial clockwork.

Let us take a very brief look at what all the fuss was about. If one could ignore the effect of the sun then the motion of the moon would be blissfully simple, it would move around the Earth in an ellipse following the laws discovered by Kepler. But Kepler's laws describe the motion of one body around another and this simple picture does not include the sun. The major effect of the sun is to make the axes of the moon's orbit rotate slowly, once every 8·85 years; at the same time the positions in the sky where this orbit intersects the plane of the Earth's orbit move slowly westward, making one complete revolution in 18·61 years. In addition there are three more rather complicated effects which are large enough to have been detected with the naked eye. One of these, called *variation*, is a wobble in the moon's motion due to the fact that during part of each month the sun and Earth pull it in the same direction and then in opposite directions. This causes a monthly wobble of 39 minutes of arc in its position in the sky. A second effect, called *evection*, is due to the effect of the sun's gravitational attraction on the ellipticity of the moon's orbit; this displaces the moon by up to 76 minutes of arc and has a period of 1 month. Finally, there is a third effect, discovered by Kepler, called *annual inequality* which is due to the fact that the sun's pull on the moon varies throughout the year because the Earth's orbit is elliptical; this displaces the moon by about 11 minutes of arc. As you might expect, this is not the whole story; there are a host of other minor effects due to the fact that the Earth is not perfectly spherical and to the attraction of the other planets on the moon.

Following Newton, the theory of the moon's motion was advanced by such eminent mathematicians as Clairaut, d'Alembert, and Euler but it was nearly 70 years before it was possible to predict its position with the accuracy required for navigation, 1 minute of arc. The first lunar tables to achieve this accuracy were calculated by Tobias Mayer in 1753, using methods put forward by the great mathematician Euler in a prize essay to the Academy in Paris in 1748. Mayer was a cartographer and the purpose of his work was to make better maps by improving the measurement of longitude on land; he was not interested in the sea and had never even seen it. Originally Mayer claimed that his lunar tables were accurate to 2 minutes of arc, but a subsequent comparison of them with extensive observations made at the Royal Observatory at Greenwich showed that they were accurate to 1 minute of arc and therefore good enough for finding longitude at sea by the method of lunars.

The positions of the stars

The publication of Mayer's lunar tables in 1753 solved the problem of predicting the moon's position, but what about the other half of the problem, predicting the positions of the stars? In the first half of the 18th century positional astronomy made very rapid progress due largely to the work of James Bradley in England and Nicolas Louis de Lacaille in France. Other important factors were the introduction of micrometer eyepieces for telescopes, which greatly improved measurements of a position, and the use of more accurate pendulum clocks for timing the transits of stars. Two new phenomena, both important to positional astronomy, were discovered at this time by Bradley. In 1727, when working with a telescope at his aunt's house in Wanstead, he discovered an important annual variation in the apparent position of stars which he called *aberration*. This remarkable effect is due to the fact that light does not travel at an infinite speed; it travels at 300000 kilometres per second, while the Earth in its orbit round the sun travels at an average speed of about 30 kilometres per second. The apparent direction of a star seen from the Earth therefore varies throughout the year as the direction of the Earth's motion changes relative to the star, much in the same way as the apparent direction of the wind in a sailing boat depends on the direction in which the boat is going. If the boat is sailing directly into the wind, or away from it, then the apparent direction of the wind is the true direction; but if the boat is sailing at an angle to the wind then the apparent direction differs from the true direction by an amount which depends in a simple way on the relative speed and direction of the wind and the boat. Thus when the Earth is travelling towards or away from a star there is no shift in the star's position, but if it is travelling at an angle to the star, then there is a shift which can be as large as 20 seconds of arc. In the course of a year the apparent position of a star may vary by about 20 seconds of arc and the details of this apparent motion depend on where the star is in the sky. This phenomenon is called the aberration of light and is obviously of considerable importance in positional astronomy.

It is interesting to note that an explanation of aberration depends upon knowing the ratio of the velocity of light to the speed of the Earth in its orbit. The first measurement of the velocity of light was made in 1675 when Roemer in Paris made his classic observations of the satellites of Jupiter (see Chapter 5). The first reasonable estimate for the Earth's speed in its orbit was made in about 1672 when Cassini in Paris organized a measurement of the distance of Mars from which, by Kepler's laws, it is possible to work out the size of the Earth's orbit. Thus in 1727 Bradley was able to give a correct quantitative explanation of his discovery. Quite apart from its importance to positional astronomy the discovery of aberration is of historical significance because it was the first direct experimental evidence that the Earth is moving round the sun.

The other effect discovered by Bradley is called *nutation*. We have already noted that the Earth's axis does not remain in a constant orientation in space but precesses slowly round in a circle, a phenomenon which was first explained by Newton in 1687. Newton predicted that this motion would not prove to be perfectly circular because the axes of the moon's orbit are not fixed in space but, rotate once around the Earth every 18·6 years. In consequence the direction of the moon's pull relative to the sun, is continually changing and the precession of the Earth's pole is therefore not perfectly circular but exhibits nutation; superimposed on the circular motion there are small waves of about 9 seconds of arc with a period of 18·6 years. Thus, to predict the position of a star with an accuracy significantly better than 1 minute of arc it is essential to correct it for nutation. The corrections for nutation were put on a sound observational basis by Bradley in a paper published in 1747.

As a result of all this work on positional astronomy the accuracy with which star positions could be fixed increased by at least ten times during the first half of the 18th century. By the time Mayer's lunar tables appeared in 1753 the positions of bright stars were known with an accuracy of about 5 seconds of arc, while the positions of thousands of stars were known to 30 seconds of arc; all the astronomical problems of finding longitude at sea were therefore solved.

The method of lunars

There was, of course, a great deal to be done before these astronomical advances could be put to practical use. At this critical stage the progress of navigation owes much to the effective co-operation between the Royal Society, the Admiralty, and the Royal Greenwich Observatory, who between them arranged that Mayer's lunar tables should be compared with observations of the moon by Bradley and given a practical test at sea by Neville Maskelyne, later to become Astronomer Royal. Maskelyne acted as an effective link between the worlds of astronomical science and practical navigation and it was he who pulled all the loose strings together and made the method of lunars into a practical system of finding longitude. In his voyage to the Barbados in 1763 Maskelyne showed that it was possible, using Mayer's lunar tables and Hadley's quadrant, to find longitude at sea to better than 1° or 60 nautical miles; the only difficulty was that it took about 4 hours to work out each result! Maskelyne appreciated how serious a drawback these lengthy calculations were and at once took steps to simplify them. On his appointment as Astronomer Royal he organized the computation of a table of lunar distances and persuaded the Admiralty to publish them in that invaluable aid to the navigator, the *Nautical Almanac*. The tables gave the angular distance of the moon from 9 bright stars and the sun for every 3 hours of Greenwich time and for every day of the year. The first edition of the *Nautical Almanac* appeared in 1766, and from that date the method of lunars was available to navigators for finding longitude at sea. Mayer died in 1762, but his heirs were awarded £3000 in recognition of his tables. Euler, who had provided the basic equations, was awarded £300.

In the hands of a good navigator the method of lunars was a satisfactory solution to the problem of finding longitude. The best witness is that expert navigator, Captain James Cook. In his journal, Cook describes it as 'a method that we have generally found may be depended upon to within half a degree, which is a degree of accuracy more than sufficient for nautical purposes'. Cook's statement is certainly supported by what he actually accomplished. Compare, for example, the voyage of the Endeavour (1768–1771) with the voyage of the Centurion under Commodore Anson (1741–1743) only 27 years earlier! Anson had no means of finding longitude and as a direct consequence his expedition suffered terribly. On his way from England to the west coast of South America he overestimated the distance which he had travelled to the west and at the first attempt failed to round Cape Horn. The weather was bad and he nearly put the whole fleet on the rocks. In the next leg of his journey he missed his destination, the island of Juan Fernandez and not knowing whether to sail east or west, spent so much time looking for it that a considerable number of his crew died of scurvy. That sort of thing just didn't happen to Cook; his voyages mark a new era, an era in which ships could find their position. Compared with earlier travels, Cook's voyages read like a railway timetable. On his first famous voyage he left England on 25 August 1768 and, via Cape Horn, dropped anchor in Tahiti on 13 April 1769, in time to carry out his objective, the observation of the transit of Venus across the face of the sun, for which he had perfect weather. He went on to make an extensive and remarkably precise survey of the coasts of New Zealand and eastern Australia, returning in good order to England via the Cape of Good Hope. Cook carried Maskelyne's *Nautical Almanac* and Hadley's quadrant with which, aided by the astronomer Mr Charles Green, at one time an assistant to Maskelyne (see p. 73), he made hundreds of measurements of longitude using the method of lunar distances.

Digressing for a moment, there is another aspect of Cook's voyage which impresses me. How did anyone persuade the King or his advisers to finance an expedition to observe the transit of Venus at a cost of £4000? The observation was designed to measure the distance of the sun by a new technique suggested by Halley in 1716. By timing the passage of Venus across the disc of the sun from two widely separated places on Earth it is possible to work out the parallax of Venus, that is to say, its distance in terms of the size of the Earth. Given that information it is simple, from Kepler's laws, to find the distance of the sun. A previous attempt to do this in 1761 had been inconclusive and, if it was not tried again in 1769, it would have been necessary to wait for more than a hundred years, until 1874, for another opportunity. No doubt the astronomers of the time thought it an important measurement, but from a strictly practical point of view it was not really urgent. The solar distance had been found by Cassini from observations of Mars in 1672 and, more validly, by Lacaille in 1757. The original proposal by the Royal Society in 1767 is therefore understandably vague from a cost–benefit point of view and says that it is 'a Phaenomenon that must, if the same be accurately observed in proper places, contribute greatly to the improvement of Astronomy on which Navigation so much depends'. But it goes on to say, 'that several of the great Powers in Europe, particularly the French, Spaniards, Dutch and Swedes are making the proper dispositions for the Observations thereof, and the Empress of Russia has given directions for having the same observed in many places in their extensive Dominions'. The final argument is that 'as the British Nation have been justly celebrated in the learned world for their knowledge of Astronomy in which they are inferior to no Nation upon Earth, Ancient or Modern, it would cast dishonour upon them should they neglect to have correct Observations made of this important Phaenomenon'. I think we may safely conclude that this is an 18th century example of 'grantsmanship' in which one exploits the national prestige of science, not its utility, to get money out of the Government. Another familiar ploy, used effectively in our own time by the space programme, is to harp on the military value of the work. I am left wondering whether the national prestige of measuring the distance to the sun was really sufficient to explain the large sums of money granted to this project.

To return to the problem of longitude, although the method of lunars solved the age-old problem of finding longitude at sea it really came too late. It had a surprisingly short life and before it could gain popularity, especially among merchant sailors, it was replaced by the second method — the use of a transportable clock.

Harrison's first timekeeper (H1) completed in 1735. It was initially tested on a barge in the Humber and then subsequently on a voyage to Lisbon.

The marine chronometer

It had always been realized that in principle the use of a transportable clock was the simplest of all methods of finding longitude but in practice the mechanical problems of developing a suitable clock took longer to solve than the mathematical problems of the method of lunars. The first practical timepiece which was sufficiently accurate to find longitude when carried in a ship was made in London by John Harrison. When the offer of the great prize for longitude was made in 1714, Harrison was 17 years old and he apparently devoted his life to winning it.

The technical problem which Harrison solved was to develop a clockwork mechanism which would run at a uniform rate in a moving ship under conditions of varying temperature. He eventually cracked this problem with his fourth timepiece (H4) which he completed in 1759 at the age of 66. Harrison made it to look like an ordinary pocket watch although it was 5 inches in diameter. At sea this enormous pocket watch was laid on a cushion in a box whose level could be adjusted by hand—Harrison distrusted gymbals. The first official test of H4 was made on a voyage to Jamaica in 1761 when, after 81 days at sea it was found to be 5 seconds slow assuming, of course, the longitude of Jamaica to be correct. At the equator this error would correspond to an error in longitude of $1\frac{1}{4}$ minutes of arc or $1\frac{1}{4}$ nautical miles and so the performance of H4 was considerably better than the $\frac{1}{2}°$ or 30 nautical miles required to win the major prize of £20 000. Nevertheless, the Board of Longitude, charged with administering the prize, argued that Harrison's success might be just a stroke of luck. They demanded a second trial and based their decision formally on the argument that the longitude of Jamaica was not known with sufficient precision to be sure of the accuracy of H4! At first sight this whole affair looks absurd, but it was not quite as silly as it looks. The original intention of the Board had been to establish the longitude of Jamaica by the classical method of observing the eclipses of the satellites of Jupiter using two observers, one in England and one in Jamaica. The longitude of Jamaica had been established some years before by an observation of the transit of Mercury across the sun's disc and there was some doubt about its accuracy. However, owing to a delay of 6 months in the start of the test, this plan could not be carried out, because Jupiter was no longer in a convenient position.

A second, more scientific test, was carried out in 1764 between England and the Barbados. Two astronomers, Maskelyne and Green, were sent ahead to measure the longitude by observing the eclipses of the first satellite of Jupiter. Harrison's 'watch' (H4) was embarked on H.M.S. *Tartar* in the charge of his son, William. On arrival in the Barbados, 7 weeks later, it was found to be 40 seconds fast, corresponding to an error in longitude of about 10 minutes of arc. This time there was no doubt about the longitude of the Barbados, which was settled beyond question by the many observations made by Maskelyne and Green. In fact the only person who raised doubts was William Harrison who, not unreasonably, objected to Maskelyne as an observer on the grounds that he was an active sponsor of the rival method of lunars. Nevertheless, the performance of H4 was verified by Maskelyne and Green and the Board agreed that it met the requirements of the major prize for longitude.

The subsequent history of the negotiations between Harrison and the Board of Longitude are long, involved, and mostly acrimonious. Following his disclosure of the mechanism of H4 he was granted half the prize in 1765, but it took him another 8 years, the personal help of George III, and an Act of Parliament to get the balance. He was then 80 years old; it had taken a lifetime to win the prize.

One of the essential conditions of the prize was that it should be possible to have it copied and this task was entrusted by the Board to another London watchmaker, Larcum Kendall. His first copy, K1, cost £450 and was given to Captain Cook to test on his second great voyage to the southern polar regions in 1772. The second copy, K2, was simplified to reduce the price and cost £200. It was issued to Captain Bligh in 1787 for his ill-fated voyage on the *Bounty*. When the crew mutined and put Bligh and his officers off in a boat they took K2 with them to Pitcairn where it stayed until 1808, when it was bought by the captain of a whaler. Again it was stolen and landed in Chile, whence it was returned to England in 1843. The third copy, K3, was made for £100 and went with Cook on his third, disastrous, voyage in 1776; it was subsequently used by George Vancouver and Matthew Flinders. But it was K1 which was most important. In his report on the extensive tests of KI Cook wrote, 'Mr Kendall's Watch has exceeded the expectations of its most Zealous advocate and by being now and then corrected by Lunar observations has been our faithful guide through all vicissitudes of climate'. No more need be said; Harrison, Kendall, and Cook had shown that a marine chronometer could be made and used as a valuable aid to navigation.

The marine chronometer gained in popularity and in the first half of the 19th century it displaced the method of lunars completely. At first it was too expensive and one must remember that ship's officers were expected to buy their own instruments. As Captain Cook wrote, 'the most expensive article, and what is in some measure necessary in order to come at the utmost accuracy, is a good watch; but for common use and where the utmost accuracy is not required, one may do without'. However, the pressure for more accurate navigation was now supported by the great volume of trade by sea and the commercial importance of precise navigation, with its attendant benefits of safety and punctuality. The simplicity of finding longitude with a chronometer was irresistible, especially to merchant seamen, and the demand grew. By the end of the 18th century we find John Arnold making and selling robust marine chronometers for 60 guineas. By 1825 the marine chronometer had well and truly arrived; in that year it was officially supplied to H.M. ships.

It is interesting to note that the introduction of the marine chronometer not only solved the problem of measuring longitude but also made possible more flexible methods of fixing position which did not involve finding latitude and longitude separately. The best example is *Sumner's method*. In his book published in 1843, Captain Charles Sumner of Boston describes how, a few years previously, he was caught in a gale off the Irish coast and was trying to find out where he was. He managed to make only one observation of the altitude of the sun, which he timed with a chronometer. Although this single observation did not tell him the position of his ship, Captain Sumner realized that it did tell him that the ship must lie somewhere on a line (actually a circle) which he could draw on his chart—a remarkable discovery to be made by the captain of a ship during a storm.

Harrison's fourth timekeeper (H4) completed in 1759. It won him the prize offered by the British ▷
Parliament in 1714 for finding longitude at sea.

A marine chronometer made in 1818 by J.R. Arnold. His father John Arnold, was one of the most famous English chronometer makers, with a factory at Chigwell (London). Between 1787–99, J.R. Arnold and his father ran the London firm of Arnold and Son, makers of pocket and marine chronometers. Later, J.R. Arnold had a shop at 102 Cornhill (1804) and at 26 Cecil Street, London (1816–30).

Following this experience he developed Sumner's method which makes use of the fact that, if a circle is drawn on the globe with its centre at the point where the sun, or any other star, is vertically overhead, then from all points on that circle the altitude of the sun is the same. Thus the navigator first measures the altitude of the sun with a sextant, noting the time at Greenwich as shown by a chronometer, and then with the aid of an almanac he works out where on the globe the sun was vertically overhead at that time. He then draws a circle round that point with a radius equal to the observed zenith angle (90° minus altitude) of the sun, and the ship must be somewhere on that circle. To fix the position of the ship he must make two observations and draw two intersecting circles. These observations can be made on two different stars, preferably about 90° apart in azimuth, or they can both be made on the sun. In the latter case it is necessary to make the two observations at a sufficient time interval for the direction of sun to have changed substantially and to take into account the movement of the ship during this interval. There are many later refinements of this way of fixing position but they are mostly based on Sumner's method.

The use of the chronometer was not confined to ships, but was also applied to improving the measurement of longitude on land. By the method of lunars or Jupiter's satellites it had been possible for many years to measure longitude on land with an accuracy of roughly 5 minutes of arc, corresponding to an uncertainty in time of 20 seconds. This was adequate in the days when the uncertainty in a ship's longitude was commonly $\frac{1}{2}°$ or 30 nautical miles, but following the introduction of the chronometer, this uncertainty was substantially reduced on all but the longest voyages. There was therefore a considerable interest in improving the accuracy of maps and also in establishing the difference in longitude between different observatories so that the measured positions of stars and also measurements of time could be compared. In the years between 1849 and 1851 the difference in longitude between Liverpool (England) and Cambridge (Massachusetts, U.S.A.) was measured with great care by the U.S. Coastal Survey. In this particular work 50 marine chronometers were sent in six separate voyages across the Atlantic. The final uncertainty in the difference in longitude was 0·039 seconds of time which, at that latitude, corresponds to a distance of about 50 metres.

Predicting the ocean tides

There is one important problem in navigation which engaged astronomers for many years and which we have barely mentioned. What is the best way of predicting the ocean tides? The tides are such a routine event that most of us take it for granted that their explanation is obvious and has something to do with gravity and the moon. And yet, although the first tide-table in England—*Flod at London Brigge*—dates from the 13th century, it is only comparatively recently that the origin of the tides has been understood. I suppose most of us picture the moon as pulling the water of the ocean towards it so that there is a heap of water, a high tide, at the place where the moon is overhead. There is a good deal of truth in this picture but, as we shall see, there is also a lot wrong with it.

The first obvious fault is that it should be the moon, not the sun, which has the major influence on the tides and yet the gravitational force due to the sun at the Earth's surface is very much greater than that of the moon. A second difficulty is that there are two tides in every day and yet our simple picture suggests that there should only be one. Finally, the highest tides occur when the sun, moon, and Earth are all in a straight line but, surprisingly, it doesn't make any difference whether the sun and moon are both on the same side of the Earth, as at new moon, or on opposite sides, as at full moon.

Newton gave the answers to all these tricky questions in 1697 and they are in the textbooks. He showed us, for the first time, how the sun and moon raise tides and why the moon has the major influence. The

77

explanation is far from obvious and, even now, is not widely understood.

Briefly, the Earth and moon orbit around their common centre of mass and are held together by their mutual gravitational attraction which continuously accelerates them towards their common centre. However, this gravitational force is not the same for all points on the Earth because it depends upon their distance from the moon. Thus at the centre of the Earth the attractive force of the moon is exactly strong enough to keep a particle in the orbit followed by the centre of the Earth. But on the side of the Earth nearest to the moon the force is slightly higher and a particle on the surface feels a small resultant force *towards* the moon; this force, 'pulling' on the water of the oceans, raises a high tide, as one would expect. On the far side of the Earth exactly the opposite happens; the attractive force of the moon is weaker than at the centre and a particle on the surface feels a small resultant force *away* from the moon. As a result the water of the oceans on the far side of the Earth is 'pushed' away from the moon and forms an equal and opposite high tide, as one would not expect.

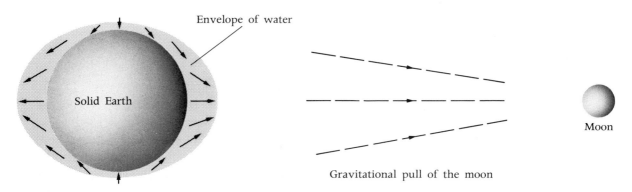

The pull of the moon on the Earth's envelope of water raises two high tides in the oceans on opposite sides of the Earth. The arrows show the direction of this tide-raising force.

Thus the forces which raise the tides are not as simple as one might suppose. They are *not* simply proportional to the gravitational force of the sun or moon, but to the *difference* between the forces on opposite sides of the Earth. Because the moon is so close this difference is relatively large; on the other hand, because the sun is so far away it exerts roughly the same force on both sides of the Earth and the difference is relatively small. Thus, although the gravitational force of the sun at the surface of the Earth is 180 times greater than that of the moon, the *tidal* force due to the sun is only half that due to the moon and it is therefore the moon which has the major effect in raising the tides. In passing, it is surprising how small these tidal forces are; the tidal force due to the moon is roughly 10 million times less than the force of the Earth's gravity and, although it produces the impressive phenomenon of the tides, it is quite difficult to detect in the laboratory.

The reason why there are two tides in every day, and not one, is now clear. There is a high tide on *both* sides of the Earth, one where the moon is directly overhead and one where it is directly beneath our feet. Thus as the Earth spins on its axis we shall see one high tide when the moon is high in the sky and another 12 hours 25 minutes later when the Earth has made half a revolution with respect to the moon and the moon is under our feet. These two tides will not necessarily appear to be equal because their relative heights depend upon how close we pass, as the Earth spins, to the points where the moon is directly overhead or directly underneath. When the moon is over the equator we shall pass at an equal distance from these two points and the two tides will appear to be equal but, generally speaking, the relative heights of two successive tides will depend in a rather complicated way on the declination of the moon and on our own latitude.

The sun produces its own tide, roughly half as big as the tide due to the moon. When the sun and moon are in line with the Earth these two tides are in step and add together to form a large tide (spring tide); when they are at right angles to each other, as seen from the Earth, the two tides are out of step and the resultant tide is small (neap tide). We can now see why spring tides occur at both full and new moon. The sun and moon produce their tides equally on both sides of the Earth and so it makes no difference whether they are on the same side of the Earth or on opposite sides.

Broadly speaking, Newton succeeded in explaining how the sun and moon raise the ocean tides and in so doing he made a major advance in our understanding of the world, but in several important respects his theory gave the wrong answers. For example, in most places high tide does not occur at the moment when the moon is on the meridian, due north or south; in fact the 'tide interval', as it is called, can be anywhere from zero up to the maximum possible value of 6 hours. Furthermore, neither the actual height of the tide nor the difference between successive tides agrees well with his theory. However, the shortcomings of Newton's theory are not astronomical but hydrodynamical. Although he explained the astronomical forces which raise the tides, he did not consider how the water actually flows. A complete theory of the tides must take into account the resistance to tidal flow offered by land masses and the speed with which tidal changes travel. Thus it involves a host of intricate details, such as the shape of the coastline and depth of the ocean, and it is only practicable to compute theoretical tides for simple and restricted regions.

Practical tide-tables are based on a judicious combination of tidal measurements with astronomical theory. The most accurate method of prediction is based on *harmonic analysis*. In this method astronomical theory is used to represent the total tide-generating force as the sum of a large number of smaller *harmonic components* of force which correspond to the major periodicities of the Earth–moon–sun system, such as the lunar day, solar day, variation of the declination of the sun and moon, and so on.

ENGLAND, WEST COAST — LIVERPOOL
Lat. 53° 25′ N.   Long. 3° 00′ W.   125

TIME ZONE: G.M.T.   TIMES AND HEIGHTS OF HIGH AND LOW WATERS   YEAR 1976

JULY								AUGUST								SEPTEMBER							
	Time	m	Ft.		Time	m	Ft.		Time	m	Ft.		Time	m	Ft.		Time	m	Ft.		Time	m	Ft.
1 Th	0113	9·2	30·1	16 F	0158	9·0	29·5	1 Su	0222	9·3	30·5	16 M	0233	8·4	27·5	1 W	0342	8·4	27·6	16 Th	0315	7·6	24·9
	0811	1·1	3·7		0846	1·4	4·7		0920	1·1	3·6		0912	2·1	6·8		1035	2·1	6·9		0952	2·8	9·3
	1342	8·9	29·1		1423	8·4	27·7		1452	8·9	29·2		1453	8·0	26·2		1617	8·2	26·8		1538	7·5	24·5
	2023	1·4	4·7		2059	1·8	5·9		2131	1·5	4·8		2129	2·3	7·6		2300	2·3	7·6		2227	3·0	9·8
2 F	0155	9·1	29·9	17 Sa	0237	8·7	28·4	2 M	0308	9·0	29·4	17 Tu	0311	8·0	26·1	2 Th	0451	7·9	25·8	17 F	0415	7·2	23·7
	0852	1·2	4·0		0919	1·8	5·9		1003	1·5	4·9		0948	2·5	8·1		1143	2·6	8·5		1051	3·2	10·4
	1426	8·7	28·7		1502	8·1	26·6		1541	8·5	28·0		1534	7·6	25·0		1729	7·8	25·7		1644	7·2	23·6
	2104	1·6	5·2		2133	2·1	7·0		2218	1·9	6·2		2212	2·7	8·9						2337	3·2	10·5
3 Sa	0240	9·0	29·4	18 Su	0317	8·3	27·1	3 Tu	0402	8·5	27·9	18 W	0358	7·5	24·6	3 F	0021	2·6	8·6	18 Sa	0531	7·0	23·1
	0935	1·4	4·6		0955	2·2	7·1		1055	2·0	6·4		1034	2·8	9·3		0615	7·6	24·9		1206	3·3	10·7
	1514	8·5	28·0		1543	7·8	25·5		1640	8·2	26·9		1627	7·3	23·9		1309	2·7	8·9		1805	7·2	23·6
	2148	1·8	5·9		2213	2·5	8·1		2316	2·3	7·5		2306	3·0	10·0		1854	7·8	25·6				
4 Su	0330	8·7	28·6	19 M	0401	7·8	25·7	4 W	0508	8·1	26·5	19 Th	0459	7·2	23·5	4 Sa	0153	2·5	8·1	19 Su	0102	3·1	10·1
	1023	1·6	5·4		1037	2·5	8·3		1202	2·3	7·6		1133	3·1	10·2		0745	7·7	25·3		0655	7·2	23·7
	1607	8·3	27·3		1631	7·4	24·4		1750	8·0	26·1		1734	7·1	23·3		1432	2·4	8·0		1330	3·0	10·0
	2238	2·0	6·7		2259	2·8	9·2										2014	8·1	26·6		1926	7·5	24·7

Part of a modern tide table for Liverpool.

The basic idea is that each harmonic component of the tide-generating force will raise its own tide. A complete list of these components is incredibly complicated, involving about 400, but a great deal of useful work has been done with only the 7 major components, all of which are diurnal or semi-diurnal. The next step is to analyse a series of observations of the height of the tide at a particular port to find the partial tide associated with each harmonic component. The factors which describe the height and time delay, or phase, of this partial tide are known as the 'harmonic constants' of the port and have been measured for thousands of ports.

The tide for any port can now be calculated for any date, past, present, or future by evaluating the harmonic components for that date from astronomical tables and then using the harmonic constants to compute the partial tides and hence the total tide. If we want to predict the tides for several years ahead, then we must, of course, include in the analysis some of the long term harmonic components. Experience has shown that this method of predicting the tides gives good results, especially in deep-water ports which are least affected by local variations in the sea-bed.

As far as astronomy is concerned, the answers to questions about the ocean tides were given years ago, first when Newton published his *Principia* in 1687 and later when the motion of the moon was understood in the 18th and 19th centuries. The answers were certainly not among the most urgent or glamorous in the history of astronomy but they were of considerable practical value and they do illustrate the close connection between pure and applied science. Thus in order to explain the tides, Newton brought to bear a wealth of information about the solar system, such as the mass and distance of the sun and moon, which interested speculative thinkers in ancient Greece but which most people, even now, would regard as useless.

Navigation in the first half of the 20th century

The art—or should one say science?—of navigation was completely transformed in the first half of the 20th century by radio and the aeroplane; to the navigator the radio engineer became a more important figure than the astronomer. From a strictly navigational point of view the first application of radio was the transmission of time signals which started in Boston in 1904. Radio time signals made Greenwich Mean Time available to ships at sea so that they could check their chronometers and, incidentally, made possible more precise measurements of the longitude of places on land. A later application was the radio direction-finder which enabled ships and, later, aircraft to take bearings on radio stations on land. However, these were both minor advances compared with what was to come when a whole new generation of radio aids were developed, largely in response to the urgent needs of military aviation, during the 1939–45 war. Up to that time most aircraft had found their way about much in the same way as ships; that is to say by compass, map reading, an occasional radio bearing and, on long flights over the sea, an occasional fix by the stars. Indeed a special sextant for aircraft was developed from the marine sextant and a special *Air Almanac* was developed from the *Nautical Almanac*. Nevertheless these methods, originally evolved for ships, were not suited to fast-moving aircraft which have to know precisely where they are at night and in cloud. In fact the further development of navigation in the air, both civil and military, relied entirely on developments in radio engineering and not in astronomy.

One of the first of these new aids was radar which dates from about 1935. As far as ships were concerned radar, like the telescope in 1608, extended the mariner's vision. In effect it allowed him to 'see' in fog or in the dark as far as he could see on a clear day. Apart from the fact that he was now aware of nearby ships, he could use the time-honoured methods of coastal navigation even under conditions of bad visibility. As far as aircraft were concerned radar also extended their vision so that they could 'see' in cloud or in the dark but, of equal importance, it made it possible for ground stations to 'see' the aircraft under all conditions of visibility. One has only to look at the problem of handling the air traffic near a large airport, especially in bad weather, to realize how essential radar is to the development of flying.

At about the same time a whole host of radio navigational aids were developed, mainly for aircraft. There are several short range systems for guiding aircraft between airports, and there are three principal long-range systems (Loran, Decca, Omega) which can be used by both ships and aircraft. These are all based on the use of large fixed transmitting stations which transmit signals which are synchronized. On the ship or aircraft a radio receiver measures the relative time of arrival of the signals from one pair of ground stations and from this information the navigator can place himself somewhere on a line, actually a hyperbola, which is marked on a specially prepared chart. He can now fix his position by getting a second, intersecting, position line from another pair of stations. These systems all have the advantage, compared with astronomical methods of fixing position, that they work under all conditions of visibility. Their accuracy cannot be stated simply because it depends upon where the receiver is in relation to the network of ground stations but, roughly speaking, it is of the order of one or two kilometres, which is comparable with the accuracy of a position found by conventional astronomical methods using modern instruments.

However, in recent years there seems to have been a return to 'celestial bodies' for navigation; perhaps the astronomer is coming back into fashion. After all, an object like the moon has the enormous advantage that it can be seen over a wide area of the Earth including the great oceans; if therefore one can make an artificial moon, shining by radio waves which penetrate cloud, then it is preferable to a widely dispersed network of radio stations on the ground which have a limited range. The first and best known of the new systems is the U.S. Navy's Navigational Satellite System (TRANSIT) which uses five earth-orbiting satellites launched between 1967 and 1970. These satellites are at low altitude, about 1000 km, and are in polar orbits; that is to say they travel over the north and south poles. Each of them carries a stable oscillator and transmits on two different radio frequencies. The ship carries a radio receiver and measures the apparent difference between these two frequencies which, due to the well-known Doppler effect, is a measure of the relative velocity of the satellite and ship. From measurements of this velocity, made over a period of several minutes, and from a knowledge of the satellite's orbit, the position of the ship can be calculated with a small computer. The accuracy of the position is roughly 200 metres and the coverage of the system is effectively worldwide. It is difficult to see what more one could want!

All these modern radio techniques depend upon very precise radio frequencies and this necessarily implies that, somewhere, there is a very precise standard of time. In Chapter 2 we have outlined the influence of this demand on our modern system of measuring time.

Inertial navigation

None of the navigational aids which we have looked at so far is perfect for all users. If they use the stars or the sun, they rely on a clear view of the sky which cannot be guaranteed. If they use radio signals, there are a variety of other drawbacks; they are limited by the location of the transmitters or by conditions of good reception; furthermore, from a military point of view, there is the very serious objection that they can be put out of action by radio interference or jamming. If we were asked to specify the perfect navigational aid I suppose we should visualize it as a small, lightweight box, which would indicate continuously and precisely its position on the Earth. Ideally such a box would work high in the sky for aircraft and under the sea for submarines; it would be self-contained and it would always 'know' where it was without the help of the sun or stars or radio stations.

It is an astonishing fact that something very like our ideal box is emerging from recent work on 'inertial navigation'. The basic tools of this new technique are the gyroscope and the accelerometer. A gyroscope is a spinning top which, as most children know, resists any attempt to change the direction in space of its axis. An accelerometer is a device for measuring acceleration along one particular direction. It consists of a small mass freely supported in such a way that when it is accelerated along the axis of the instrument, the force causing the acceleration can be measured. Newton's second law of motion tells us that if we divide this force by the mass we find the acceleration.

The gyroscope and accelerometer can be used in a variety of ways to produce a navigational aid and the oldest application, the gyro-compass, has been used in ships for over 50 years. However it is only in the last few years that improvements in the gyroscope have made it possible to combine it with accelerometers to produce a whole new generation of 'inertial' navigational aids which promise to displace the radio aids which, in their turn, took the place of navigation by the sun and stars. Let us look very briefly at one way in which our ideal box might be made. The basic idea would be to measure the acceleration of the box in three directions in space which are mutually at right angles and are fixed relative to the Earth; for example, north–south, east–west and up–down. By integrating the product of these accelerations and the time for which they are observed we can find the three velocities of the box, and again by integrating the product of these velocities and times, we can find the distances which the box has travelled in the three directions. If, therefore, we know the position of the box at the start, we can find out its position at all subsequent times.

One way in which this idea can be put into practice is to mount three accelerometers, with their axes at right angles to each other, on a 'stable platform'. The platform is kept level and in the correct orientation by two gyroscopes mounted north–south and east–west. Such a device, coupled to a suitable computer and clock, can indicate its position continuously and, like our ideal navigational aid, is entirely self-contained. Similar devices, in various forms, have been built in recent years for ships and have been used, for example, to guide submarines under the polar ice. They have also been developed for aircraft and are currently in regular use as long-range navigational aids on some of the larger civil aircraft; they have, of course, also been used on spacecraft and guided missiles. At the present time they fall short of our ideal navigational aids in accuracy, size, weight, and cost. The principal limitation to their accuracy is the difficulty of making a gyroscope which will remain perfectly fixed in direction; all practical gyroscopes wander slowly, thereby introducing an error in the indicated position which increases with time. Even so, a modern inertial aid gives the position of an aircraft to a kilometre or so after a journey of a few hours and is therefore comparable in accuracy with other long range methods of navigation. In the case of the more elaborate equipment developed for ships and submarines this sort of accuracy can be maintained over a period of several days.

Looking to the future it seems likely that inertial aids will approximate closer and closer to our ideal navigational aid and that, especially for the long-range navigation of aircraft, they will be used in preference to radio aids. Perhaps after centuries of experiment we have at last found the best technique for helping us to find our way about the Earth and, maybe, about the space around it.

Space navigation

As we have already seen, the navigational difficulties of exploring the Earth and of trading with distant lands, not to mention naval warfare, acted as a powerful stimulus to the development of astronomy. From the 15th to the 18th centuries the progress of astronomy and navigation went hand in hand until, in the present century, navigation became almost a branch of radio engineering. However, once again we are exploring, not the Earth but the solar system, and in space there are new navigational problems for the astronomer.

In finding our way about the Earth we had to develop methods of fixing position on the surface of a sphere. In space we must fix the positions of planets and space ships in three dimensions; furthermore the planets are moving. Instead of ocean currents and winds we must now reckon, as every reader of space fiction knows, with gravitational fields.

Broadly speaking, the first job for the astronomer is to make a much more accurate and detailed map of the solar system and for this we must turn to the classical study of celestial mechanics. At the beginning of the present century this study was almost entirely concerned with the problems of the solar system and had reached a most unattractive stage, dependent for further progress on thousands of precise optical observations and a morass of tedious calculations. Furthermore, there didn't seem to be much reward for knowing the motions of the planets more accurately. In fact, the study of celestial mechanics dominated the teaching of astronomy at that time and was so dull that many bright students turned to the more promising subject of physics. But now the whole scene has been transformed by radar and the electronic computer and by the urgent needs, and money, of space research. Radar allows us to measure two new parameters, range and velocity, which have proved to be of the greatest value. The computer has removed the drudgery of calculation and has made it possible to make prodigious computations with a speed and precision which have revitalized research into celestial mechanics.

Mapping the solar system

One of the first things which has to be done is to measure much more precisely the distance scale of the solar system. How far is the sun from the Earth in kilometres? If one is going to send space ships to the planets it is obviously essential to know precisely where they are. The *relative* distances of the planets from the sun have been known for many years with high precision; our problem is to turn this information into *absolute* distances without losing this precision. Kepler's third law of planetary motion tells us that the square of the time which a planet takes to go around the sun is proportional to the cube of its average distance from the sun. Thus, it is simple to find the relative distances of the planets by timing them as they go around the sun, but to find them absolutely we must measure at least one of these distances in kilometres. It can be the distance from the Earth to the sun, or, if we prefer, from the Earth to any other planet.

Attempts to measure the size of the solar system have a very long history going back at least to ancient Greece, but the results were most uncertain. Since then astronomers have tried to improve the precision and we have already seen that it was in order to measure the distance of the sun that Captain Cook went to Tahiti in 1769 to observe the transit of Venus. The results of that expedition, combined with the other observations taken at the same time, did in fact improve our knowledge of the distance to the sun but the uncertainty was still 1 part in 200 or 0·5 per cent. Such an accuracy is not good enough for space navigation, but those 18th century measurements were nevertheless only about ten times less accurate than the best modern optical measurements which were made in 1930. In that year the minor planet Eros—a most unromantic lump of rock about 20 km across—approached within 22 million kilometres of the Earth which, by planetary standards, is fairly close. It was photographed against the background of stars by many observatories and its distance from the Earth was found by the classical methods of triangulation. It was claimed for these measurements that they gave the distance to the sun to an accuracy of 1 part in 20000 or 0·005 per cent. However, there must have been something wrong which no one spotted, because we now know that they gave a distance which was too short by nearly 10 times the accuracy claimed.

The real breakthrough in surveying the solar system came after World

Astronaut Edwin E. Aldrin walking towards the laser ranging retro-reflector and the lunar module during the Apollo 11 expedition to the moon in 1969.

War II when radio techniques were applied to astronomy. In fact radio made such an impact that a substantial part of astronomy, like navigation, was taken over by the radio engineers and a new species appeared, radio-astronomers. However, the techniques of radio are now being absorbed into the mainstream of astronomy, like optics in earlier days, and radio-astronomers are slowly becoming plain astronomers. Anyway, it was the application of the wartime techniques of radar to astronomy that in 1961 gave us a precise measurement of the distance from the Earth to Venus and hence to the sun. Strictly speaking, radar measures time and not distance; it measures the time which it takes a radio wave to travel to the target and back and, in order to convert this time into kilometres, we must know the velocity of the wave which, in empty space, is the same as the velocity of light. The result of these radar observations, confirmed by eight different observatories, gave the average distance from the Earth to the sun as 149 598 500 km with an uncertainty of 500 km. Thus, the scale of the solar system is now known to about 1 part in 300 000 (0·0003 per cent) which is the sort of accuracy needed for space navigation. If we are planning to land on the nearest planet, at least 40 million km away, we can now be sure of the distance to the nearest 100 km.

The journey to the moon is an entirely different matter. Unlike the planets the distance has to be measured separately because it depends on the size of the orbit of the moon around the Earth, not around the sun, and therefore cannot be found from the other distances in the solar system.

Nevertheless it is comparatively easy to measure and has been known from optical measurements for many years with an accuracy of roughly 2 km or roughly 1 part in 200 000. In the last 20 years this uncertainty has been even further reduced by radar to less than 1 km and the precision was then limited only by the roughness of the moon's surface. Even more recently a special optical reflector was installed on the moon by the astronauts and since 1969 the distance has been monitored by transmitting short, intense, pulses of light from a laser and receiving the reflected pulses with a large optical telescope. Using the 107-inch telescope at the McDonald Observatory in Texas it has been possible to measure the travel time of the pulses to the nearest 2 nanoseconds (2×10^{-9}s), corresponding to a distance of about 60 cm. This doesn't necessarily mean that we know the absolute distance of the moon to an accuracy of 60 cm in 384 000 km because there are uncertainties in the velocity of light, but it does mean that we know it accurately enough.

To make a map of the solar system there are other things which we need to know. We must be able to predict the position of a planet in space at any given instant and to do this mathematical models of the solar system have been developed using very large and fast computers. These models involve a stupendous number of calculations which, starting from an initial configuration of the planets, find their subsequent positions, taking into account all their orbital parameters and mutual gravitational attractions. As time goes by, the facts about the solar system which are fed into these models are progressively refined by comparing them with observations. As an example, a comparison of the predicted ranges and relative velocities of Venus and Mercury with several years of radar observations has recently been made, and has yielded remarkably precise estimates of the masses of the four inner planets and of the radii of Mercury and Venus.

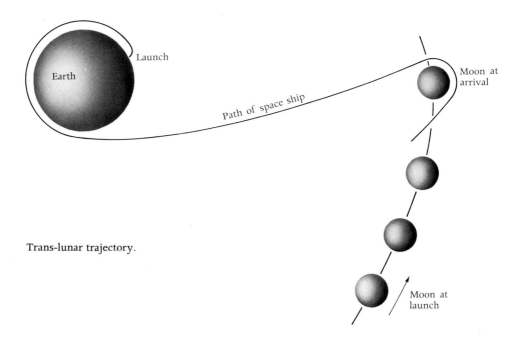

Trans-lunar trajectory.

If we plan to land on a planet then there are, of course, a whole host of things we must find out, such as the speed and direction of rotation, atmospheric composition and pressure, temperature, etc. The inner planets are already being explored and we are rapidly learning what they are like. Several men have landed on the moon and returned safely, and unmanned spacecraft have already landed on Venus and Mars and survived long enough to send us back some data about their inhospitable environments.

Fixing a spaceship

The celestial navigation of a spaceship can be carried out as it is at sea, by measuring the directions of stars and planets using a special form of sextant and a source of accurate time. But there are interesting differences between navigating in two and three dimensions. For example, if the navigator of a ship at sea measures the altitude of a star and looks at his chronometer then, as Captain Sumner realized in 1843, he knows that his ship lies somewhere on a circle which he can draw on a chart. If now he measures the altitude of a second star, he can draw a second circle on the chart and his ship lies at their intersection. If, however, the navigator of a space ship does much the same thing the result is not so satisfactory. If he uses his sextant to measure the angle between a star and some well-defined point on a nearby planet, say the centre or the edge, then he knows that he is somewhere on the surface of a cone; the apex of this cone is at the planet, it points at the star, and the half-angle at the vertex is equal to the angle he measured. If he measures the angle between the planet and a second star he gains the information that he is on the surface of a second cone. He now knows that the space ship must lie on the intersection of the two cones which, unfortunately, is not a point, but two straight lines. To fix his position in three dimensions the space navigator must obviously make another measurement and, at first sight, one would think that he should measure the angle between the planet and a third star. But rather surprisingly, this is not so; the information from a third star still leaves the space ship anywhere along a line in space. To fix the position completely the simplest thing the navigator can do is to measure the apparent angular size of the planet—the angle subtended by the whole diameter of the planet at his eye—and from this he can tell how far away it is. As he already knows that he lies on one of two radial lines from the planet, a measurement of this distance fixes two possible positions in space. There are many other ways in which he can make this additional measurement. For example, he can measure the angle between the reference planet and a nearby planet or he can measure the angle between a star and another nearby planet; the latter method gives him a fix at the intersection of three cones. The intersection of three cones in space is a peculiarly difficult thing to visualize, two are bad enough, and it goes without saying that all these navigational operations in space are aided by computers.

There are many other ways in which celestial navigation has been used in space. For example, observations with a sextant were used to check orbits of the Earth in the Gemini spacecraft and in orbiting the moon; also star sightings were used extensively on the Apollo mission to check the alignment of the inertial navigation unit. No doubt, as space exploration develops, the use of celestial navigation will develop to meet its special needs; maybe it will throw up a challenge to astronomy comparable with the classical problem of finding longitude. In the meantime, astronomy and space navigation are closely linked and are likely to stay that way; the stars are a more reliable guide to the navigator in space than on Earth, there are no clouds to blot them out when they are most needed. Perhaps the stars are no longer needed to guide our ships on Earth, but it looks as though they are still needed to guide them in space.

4 Stars and Science

Quite close to the Department of Astronomy in the University of Sydney there are Departments of Astrophysics, Geophysics, and plain Physics. Faced with these labels I suppose most visitors know that physics is the study of the properties of matter, but some might be uncertain about the need to separate geophysics and astrophysics. Does it perhaps imply that the physical properties of matter on Earth are different from those on the stars? The answer is emphatically no, these names represent nothing more than convenient labels for specialized branches of physical sciences. Indeed it is one of the basic assumptions of modern science that the same laws of physics hold everywhere in the universe. This has not always been so and a mediaeval visitor might well take the names of our departments at their face value; we must remember that to the mediaeval mind Heaven and Earth were made of different stuff which obeyed different physical laws.

In the mediaeval universe the Earth is stationary at the centre of a system of concentric and transparent spheres. Ordinary matter, the stuff of which everything around us on Earth is made, is only to be found below the moon in the first sphere, and in the technical jargon of those days was called *sublunary*. This ordinary matter is itself hierarchical and is compounded of four basic elements, fire, air, water and earth in descending order of nobility. The regions above the moon—the *superlunary* regions—are made of a more perfect type of matter called *aether*, a fifth element or essence from which, by the way, we get our word quintessence.

The laws of physics, following Aristotle, are not the same for sublunary and superlunary matter. Each of the four elements of sublunary matter has its proper place in the scheme of things where it is at rest and, if it is displaced, it tries to return home. The most humble element, earth, is at rest in the lowest place, the centre of the Earth; water seeks to be above earth, air seeks to be above water, and fire seeks to be above air. As a consequence, earth and water seek the lower regions and have a tendency to fall vertically towards the centre of the Earth, while air and fire tend to rise. The fifth element, aether, is already in its proper sphere, above the fire, and therefore has no tendency to rise or fall.

This natural tendency of the four sublunary elements to reach their respective goals explains most of the vertical motions which we see on Earth. Horizontal motions, on the other hand, can only be the result of an

Geocentric picture of the universe showing Aristotle's four elements, Earth, Water, Air, and Fire, surrounded by the spheres of the planets and the sphere of the fixed stars.

applied force; a body moves if it is pushed or pulled and its speed is proportional to the force and inversely proportional to the resistance which the body encounters. The essential point is that, if the force is removed, the body stops.

In contrast the laws of motion of superlunary matter, aether, are different. Because it is already in its appropriate region, above the moon, it has no tendency to rise or fall or, in other words, it is weightless. It can, however, move sideways, its natural motion being to circle the Earth forever at a uniform speed under the influence of some celestial prime mover.

There was another major difference between earthly and heavenly matter. It was believed that sublunary matter is corruptible, subject to

Archimedes standing on the Earth (which is flat) surrounded by water, air, fire, and the celestial spheres.

change and decay; while superlunary matter is changeless and incorruptible. Indeed this doctrine was effectively exploited by St Paul in his first letter to the Corinthians in which he uses it to answer the rather tricky physical problems involved in the Resurrection of the Dead. 'The Dead', he tells us, 'are sown in corruption', but they will be 'raised in incorruption' making use, no doubt, of the physical properties of the aether.

However, these metaphysical ideas, together with the Earth-centred universe and Aristotle's theories about motion and gravity, were questioned in the 16th and discarded in the 17th century. As we shall see in the next chapter, the final blow to the old picture of the universe came in 1687 when Sir Isaac Newton published the *Principia* in which he showed beyond reasonable doubt that the same physical laws of motion and gravity govern both the solar system and things that we can touch and see on Earth. By so doing, he removed an old barrier between astronomy and physics. Since that time great advances have been made in both sciences, and in the present chapter we shall look at some of the more important advances in physics which have been made with the help of astronomy.

Newton's laws of motion and gravity

Newton's Laws are usually taught without any historical introduction and are presented either as being obvious or as revealed truths, rather like a scientific appendix to the Ten Commandments. And yet to treat them as a revelation is to oversimplify history as Pope did when he wrote, 'Nature and Nature's laws lay hid by night, God said, "Let Newton be"—and all was light'. In fact, as Newton himself pointed out, he stood on other men's shoulders, men such as Kepler, Galileo, Descartes, Huygens, and many others. To treat the laws as being obvious is equally far from the truth. At the time they were published they were rejected by many scientists and, as any teacher knows, there are many people who still find them hard to believe; in mediaeval times they would have been regarded as wildly improbable.

The first law of motion, sometimes called the Law of Inertia, is the simplest and most basic of the three laws and it is the one which has given the most trouble. It tells us that, 'a body continues in a state of rest or of *uniform motion in a straight line*, unless it is compelled to change that state by forces impressed on it'. It is not difficult to see why this law has been so troublesome; it flatly contradicts common sense. We all know that, if a horse stops pulling a cart, the cart stops. However this common sense teaching of Aristotle, that a thing only moves if it is pushed, ran into trouble long before the time of Newton. For one thing it failed to give satisfactory answers to questions such as, 'How does an arrow fly through the air when there is nothing to push it?' Indeed the answers to this particular question gave rise to an entertaining controversy. The Aristotelians said that the air in front of the arrow rushes round to the back to fill up the vacuum, and gives it a push from behind. However this ingenious explanation did not satisfy the critics; they pointed out that, if this explanation is true, a thread tied to the back of the arrow should be blown forward; furthermore an arrow pointed at both ends should not fly so well as one pointed only at the front. I am not sure whether any of these objections were actually put to the test; probably not, for in those days they preferred argument to experiment.

The way was prepared for Newton's first law in the 14th century, when the theory of *impetus* was put forward. According to this theory a moving body acquires impetus which is carried inside it, like heat, and keeps it going. Some people said that this impetus leaks away, again like heat; others said that it is only lost by the effort of overcoming resistance to motion. The historical importance of the theory is that it accustomed people to thinking about a body as moving without anything pushing it.

The first complete and correct statement of the first law of motion was made, not by Newton, but by Réne Descartes in his *Principles of Philosophy* published in 1644, 50 years before Newton published his *Principia*.

Descartes arrived at the correct result, not by experiment but by developing the metaphysical idea that motion is a manifestation of the nature of God. The ideas of Descartes were influential for the best part of a century, largely because he developed them into a complete and impressive explanation of the whole of Creation. Before Descartes, Galileo had also arrived at the idea that a body would continue in uniform motion unless acted on by a force, but he failed to state Newton's first law because, following Aristotle, he accepted the old metaphysical idea that perfect motion is circular.

Two more laws of motion were needed to complete the picture. The second law tells us precisely what effect an external force has on the motion of a body: it accelerates it by an amount which is directly proportional to the force and inversely proportional to the mass. It is usually abbreviated to the well-worn statement 'force = mass × acceleration' and, strictly speaking, it makes the first law unnecessary. The third law tells us that 'to every action there is an equal and opposite reaction'. At first sight this looks like a rather arid statement and I am irresistibly reminded of an answer to an examination question which illustrates some of the difficulties in understanding this law. The question was, 'If Newton's third law is true, how can a horse pull a cart?'. One of the answers which illustrates the confusion which is often produced by this enigmatic law, was, 'It is true that action and reaction are equal and opposite, but the horse pulls first'! Nevertheless the third law is not so sterile as it looks—for one thing it is vital to any discussion of the important principle of the conservation of momentum.

Finally there is Newton's most famous contribution to science, his Theory of Gravity or, as it is usually known, his Law of Universal Gravitation. Newton began the *Principia* by simply stating his three laws of motion as axioms and then proceeded to show that all the motions which we observe on Earth and in the solar system can be explained if there is one more basic law, the Law of Universal Gravitation. The analysis on which this demonstration rests is extraordinarily impressive, especially when we recall that Newton was reluctant to put his ideas down on paper, and had to be persuaded to write the *Principia* by his friend Edmond Halley, later Astronomer Royal.

To prove his thesis Newton discussed in detail all the principal types of motion, including accelerated motion in a straight line and orbital motion in a circle, parabola, hyperbola, and ellipse. He showed, for the first time ever, that orbital motion in a parabola, hyperbola, or ellipse requires that there be an attractive force between the moving body and the focus of its orbit which is inversely proportional to the square of its distance from the focus. In fact he showed that all Kepler's strange laws of planetary motion are a necessary consequence of four laws, the three laws of motion and a fourth law which tells us that 'any two bodies attract each other with a force which

The flight of a cannon ball according to mediaeval physics. This Aristotelian concept of the path of a projectile was based on the belief that no body could undertake more than one motion at a time. The path therefore had to consist of two separate motions in straight lines.

is proportional to their masses and inversely proportional to the square of their distance apart'.

However, the whole of this analysis was carried out on the assumption that the moving body is a particle of *negligible* size and, before he could apply his law of gravitation to the bodies of the solar system, Newton had to show what happens when the bodies are large. The next step was therefore to show that the attraction between two spheres, no matter how large, takes place as though the whole of their mass was concentrated at their centres and that therefore all his proofs about the orbits of particles apply equally well to the solar system. He was then able to show how the Earth and the other planets are held in their elliptical orbits around the sun by the gravitational attraction of the sun and how the moon is held by the gravitational attraction of the Earth. Finally he showed how the same law of gravitation applies to the attraction between a body on the surface of the Earth and the whole Earth, and accounts for its weight and acceleration in free fall.

To show that his laws of motion and gravity explain all the principal types of motion which we observe on Earth *and* in the sky, Newton ranged over a prodigious variety of topics. The *Principia* includes a discussion of the ocean tides, the motions of the moon, the precession of the equinoxes, the oscillations of a pendulum, the orbits of comets, and many other subjects. It contains the first valid explanation of the ocean tides (see Chapter 3). It tells us how to calculate the orbit of a comet from three observations and explains quantitatively the precession of the equinoxes. Having found the key to both terrestrial and celestial mechanics it must have been difficult for a man of Newton's imagination and ability to know where to stop; everywhere he turned there were novel and fundamental contributions to be made.

As we have noted before, one of the many important contributions which the publication of the *Principia* made to science was to complete the demonstration, started by Copernicus, that the laws of physics are not peculiar to the Earth. Admittedly the dominant philosophy at that time was Cartesianism which taught that the laws of physics are universal; but, in the long run, Cartesianism carried little weight because many of its principal physical ideas were, as Newton pointed out, completely wrong. In fact, following the publication of the *Principia*, the whole structure of Cartesianism, like the teachings of Aristotle, was superseded and within 50 years was swept away. Taking a narrower view, the *Principia* was important to physics because it was the first coherent statement of the laws of dynamics and it was cast in a mathematical form which set the future pattern for physics. Taken one at a time, Newton's laws were not new; with the possible exception of the third law, they had all been put forward before. What Newton did was to choose a few simple, but correct, laws, from the many suggestions which had been made, and show how they could be made to explain everything that was then known about motion.

The relevance of astronomy to this story scarcely needs to be underlined. From the point of view of the physicist the astronomer's universe is a place where the behaviour of matter can be studied under conditions, not found on Earth, and this seems to have been true of Newton's laws. As we noted before, our experience on Earth does not prompt us to the conclusion that bodies continue in uniform motion unless they are acted upon by a force; the arrow falls to the ground, and the cart stops when it is not pulled. The fact that an apple falls to the ground suggests that it is attracted by the Earth, but nothing we see on Earth suggests that *all* bodies attract each other. Nothing on Earth points the way to the idea that gravitation is universal. Scientific discoveries are often obvious in retrospect, but to connect the fall of an apple with the idea that everything attracts everything else and so to explain the motion of the moon, is an act of genius. In the end it was the moon and planets, sailing effortlessly and eternally through space, which were the paradigms of motion. Astronomy showed physics the right way to think about the arrow and the cart.

The speed of light

If I was asked to illustrate the history of science by attempts to answer one particular question, I would choose 'What is light?'. For one thing, it would be comparatively easy to do and, for another, it is a good story; there is hardly a dull moment from the time when God said, 'Let there be light', to the present day. Astronomy has been closely involved in this story ever since it began. Let us look at one important contribution which it made, the first evidence that light travels at a finite speed.

Nowadays the speed of light is regarded by physicists as one of the most important facts about nature, but the idea that it takes time for light to travel from one place to another was not generally accepted before the 18th century. To the chief mentor of mediaeval science, Aristotle (350 B.C.), light was not a stream of particles or a wave motion in some medium but a *state* which an object acquires instantly in the presence of a luminous source. In his view it was wrong to talk about light 'travelling' and in *De Anima* he pointed out that we do not see light travel across the sky from the east to the west horizon. Almost exactly 2000 years later, Descartes taught much the same thing. It was a cardinal feature of his philosophy, Cartesianism, that there is no such thing as a vacuum, and that all the gaps between bodies are filled with an all-pervading medium, the aether. Light is propagated through this medium as a mechanical force and, just as when one end of a stick is moved the far end moves instantly, so light is propagated through the aether instantly; at one end the source pushes the aether, and at the other end this push is received instantaneously as light. There were, of course, other theories, some of which could be reconciled with the idea that light travels at a finite speed. An important one was Platonism, which taught that light is a stream of minute corpuscles. Clearly what was needed was experimental evidence.

As we might expect, Galileo tried to measure the speed of light. In his *Dialogues Concerning Two New Sciences*, he describes an experiment in which two people with lanterns take up positions some distance apart. One of them uncovers his lantern and the other, seeing the light, uncovers his in reply. The first one then notes how long it takes before he sees the reply. Galileo reports that he tried this experiment at a short distance, less than a mile, and was unable to measure any delay.

Bearing in mind that the minimum time interval which could have been measured with any certainty in the 16th century cannot have been much less than $\frac{1}{10}$th second, during which light travels about 30 000 km, it is clear that any attempt to measure the speed of light on Earth was doomed to failure. The only hope lay in using the vast distances of the solar system and that is just what the Danish astronomer Olaf Roemer did in Paris in 1676. At that time numerous observations of the eclipses of Jupiter's first satellite Io had been made because, as we noted in Chapter 3, they were used to determine longitude. These measurements showed a curious effect—the intervals between successive eclipses varied slightly about their average value of $42\frac{1}{2}$ hours. When the Earth was approaching Jupiter the intervals were about 15 seconds too short; when the Earth was receding from Jupiter they were about 15 seconds too long. Roemer interpreted this effect as being due to the finite velocity of light and, as a demonstration, predicted the correct time of the eclipse on 16 November 1676, from observations carried out in August of the same year. He estimated that the speed of light is about 214 000 km per second which compares, not unreasonably, with the modern value of 299 793 km per second; the discrepancy was partly due to an overestimate of the change in the interval between eclipses, and partly to an incorrect value for the diameter of the Earth's orbit.

Roemer's result got a fairly cool reception at the time. This is not surprising when one realizes that he was saying that light travels very roughly a million times faster than sound. Admittedly it was known that the lightning flash is seen before the thunder clap is heard; nevertheless, for those who were prepared to believe that light travels at a finite speed, it must have been hard to accept such a staggering result. We must also

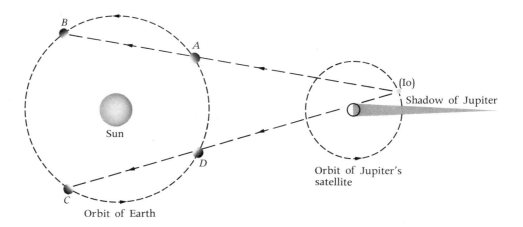

The first satellite of Jupiter (Io) orbits Jupiter in $42\frac{1}{2}$ hours. Roemer measured the speed of light by observing the time interval between the moments when Io was eclipsed by Jupiter, at different times of the year. When the Earth moves away from Jupiter (A to B) the interval is longer because the light from Io has to catch up with the Earth. When the Earth is approaching Jupiter (C to D) the time interval is shorter.

The aberration of star-light. The velocity of the Earth in its orbit around the sun combines with the speed of the light from a star to change the apparent directions of the star. This change is very small ($20\frac{1}{2}$ seconds of arc at the most) and depends upon the directions in which the Earth is moving relative to the star. In the diagram the angle of aberration is enormously exaggerated.

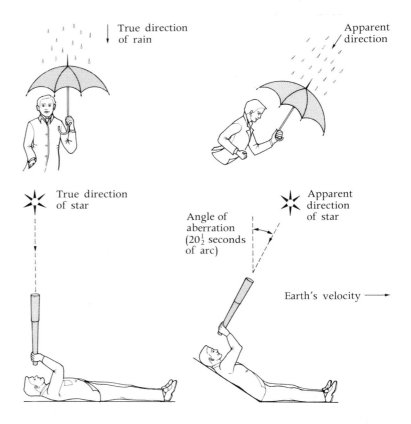

remember that the topic was of great importance to some of the arguments which were going on at the time. For example, in 1634, Descartes wrote that, if the finite speed of light could be proved the whole of his philosophy would be 'shaken to its foundations'.

What was really needed was some confirmation of Roemer's result by another method and this had to wait until 1726 when James Bradley discovered the aberration of light. Bradley observed that the positions of stars show an annual variation of about 20 seconds of arc about their mean and, as we have already noted in Chapter 3, he was able to explain this remarkable effect by the way in which the velocity of the Earth in its orbit and the velocity of the light from the star combine. Quite simply Bradley's results showed that the speed of light is about 10 000 times the speed of the Earth round the sun. They confirmed that Roemer's result was roughly correct, gave a more accurate value for the speed of light, and established, for all time, the basic fact that light travels at a finite speed.

Thus the work of two astronomers, Roemer and Bradley, gave to physics one of its central facts, the finite speed of light. For over 100 years the theory of light was based on their work alone until in 1849 Hippolyte Fizeau succeeded in measuring the speed of light over a baseline of about 9 km near Paris; his results confirmed that Bradley and Roemer were right.

The search for an aether

Another major enquiry about the world around us, in which physics and astronomy were close partners, was the search for an aether. This is a long story and to cut it short we shall pass over centuries of metaphysical speculation and take up the threads in comparatively recent times, the middle of the 19th century. At that time the question was not so much, 'Is there any aether?', as 'What properties must an aether have to do all that is required of it?' It is almost true to say that the theoretical physics of the 19th century was the physics of the aether.

A 19th century physicist needed the aether to explain the nature of light. In the previous century, largely under the influence of the followers of Newton, the theory that light is a stream of corpuscles had reigned supreme and yet, by about 1830, it had been defeated by its rival, the theory that light is a wave. This was largely due to the work of Thomas Young and Augustin Fresnel. Young was a physician in England and Fresnel was an engineer in France who turned to the study of optics after losing his job because he took part in an attempt to frustrate the return of Napoleon from Elba. Their work demonstrated, in the face of determined, and at times scandalous, opposition from supporters of the corpuscular theory that the phenomena of *diffraction* and *interference* imply that light must be a wave.

Let us digress for a moment to see what this means. In the phenomenon of *diffraction* light is observed to bend round obstacles and spread out in a way which strongly suggests that it is a wave. Historically one of the most convincing demonstrations of this was 'Arago's bright spot'. When in 1818 the Paris Academy offered a prize for an essay on diffraction, the winning entry was submitted by Fresnel, and this essay went a long way towards

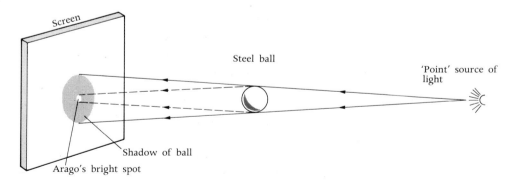

Arago's bright spot. A steel ball is hung by a fine thread in the light from a 'point' source of light. In the centre of the shadow there is a bright spot caused by the rays of light (shown as broken lines), which bend round the edge of the ball and meet in the centre of the shadow. Because they have travelled exactly the same distance they add to give a bright spot in the centre of the shadow.

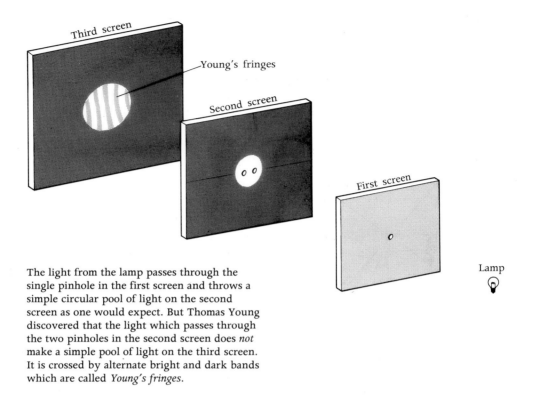

The light from the lamp passes through the single pinhole in the first screen and throws a simple circular pool of light on the second screen as one would expect. But Thomas Young discovered that the light which passes through the two pinholes in the second screen does *not* make a simple pool of light on the third screen. It is crossed by alternate bright and dark bands which are called *Young's fringes*.

establishing the wave theory of light. The eminent mathematician, Siméon Poisson, pointed out, however, that if Fresnel's theory was true, then there should be a bright spot of light at the centre of the shadow of a circular disc where the waves bending round the edge of the disc would all meet in phase with each other. On the face of it such a conclusion seemed ridiculous, until one of Fresnel's supporters, Dominique Arago, showed that the bright spot is really there! For anyone who doubts this, it is simple enough to hang a ball-bearing in a parallel beam of light and look at its shadow.

The other phenomenon, interference, is perhaps a more striking proof that light behaves as a wave. In 1803 Thomas Young showed that, if a parallel beam of light falls on an opaque screen with two pinholes in it, there is something unexpected about the light that gets through. Where the light from the two pinholes overlaps there are alternate bright and dark bands which we call *Young's fringes*. Young explained that the bright bands are in places where the light waves from the two pinholes are in step (in phase) and therefore add together to give more light; the dark bands are in places where the light waves from the two pinholes are completely out of step (in antiphase) and therefore cancel each other to give darkness. This phenomenon is called interference and can be understood in complete detail on the assumption that light is a wave. On the other hand, the fact that two light waves can cancel each other and produce darkness was, to

101

say the least, an awkward fact to explain in terms of corpuscles. And so the corpuscular theory of light lost the battle and the wave theory of light dominated the scene throughout the second half of the 19th century.   This compelling evidence that light is a wave revived a flagging interest in the aether. Quite simply the question was asked, 'If light is a wave motion, what is it that waves?' It was realized that light is not a wave-motion in air, like sound. The most obvious difference is that light travels about one million times faster than sound but there is a more basic difference. Sound waves are *longitudinal*, that it is to say the molecules of air vibrate in the same direction as the wave is going. Light, on the other hand, is a transverse wave; any medium which supports it must vibrate at right angles to the direction of the wave and for air that is a mechanical impossibility. Quite apart from all this, it was only necessary to look up at the sun or the stars to realize that light can get to us without the support of air; it had long been realized that the Earth's atmosphere does not extend far into space.

It had been known since the days of Roemer that light takes roughly 8 minutes to reach us from the sun and so, at any given moment, there must be 8 minutes' worth of energy stored in space. If there is no air, and light is a wave-motion, not a substance, where is this energy stored? There seemed to be only one possible answer—the whole of space must be filled with aether. In the jargon of the 19th century, a *luminiferous* aether was needed to support *undulatory* light. As the Marquis of Salisbury remarked in 1894, 'the main, if not the only, function of the aether has been to furnish a nominative case for the verb "to undulate"'.

What was this aether like? To a 19th century physicist it was imperative that the answer to such a question should be *mechanical*, in terms of force, mass, and physical configuration. And so it was that the mechanical properties of the aether occupied the physicists of the 19th century. Men like Lord Kelvin, James Clerk Maxwell, Leonhard Euler, and many others devoted an astonishing amount of effort and talent to this problem and they failed completely. The trouble was that no single model of the aether would explain all the things it had to do; it had to allow the planets to pass freely through it, it had to transmit the light from distant stars without too much absorption and also the flux of energy from the sun, it had to support electric and magnetic fields and gravitational attraction, and it had to explain the many complicated phenomena of optics.

If theory failed, what about experiment? To answer this question physicists turned to astronomy and visualized the Earth as sailing through an ocean of aether in its orbit around the sun. As Thomas Young said, 'I am disposed to believe that the luminiferous aether pervades the substance of all material bodies with little or no resistance, as freely perhaps as the wind passes through a grove of trees'.

Many people looked for this aether wind and the idea behind their experiments was very simple. If the Earth is travelling round the sun at 30 km per second through a space filled with aether, then there should be a gale of aether passing us; if light is a wave-motion in this aether, then the apparent speed and angle of a light beam as seen by an observer on Earth should depend on the direction of this beam relative to the motion of the Earth. A light beam would be like a boat on a fast-flowing river, so that its speed and direction as seen by someone standing on the bank would depend on the direction of the current.

The best-known of all those who tried to detect the aether wind was the American physicist Albert Michelson. His first attempt was made in Potsdam in 1881 and was financed by none other than Alexander Graham Bell, the inventor of the telephone; his last attempt was made in Pasadena in 1928. His most famous attempt, known as the Michelson–Morley experiment, was made together with Edward Morley at Western Reserve University (U.S.A.) in 1887. The essential point about Michelson's work was that he had invented an extremely sensitive technique for comparing the speeds of two light beams. He developed an instrument called an interferometer, in which a beam of light was split into two parts which were sent horizontally in two directions at right angles. After travelling 11 metres the beams were reflected back to the start where any difference between their speeds was measured by combining them in an interference pattern. The whole apparatus was mounted on a stone slab floating in mercury and could be turned so as to compare the speeds of the two beams in different directions relative to the Earth's orbital motion around the sun. Calculations showed that this motion should retard one beam with respect to the other by 0·4 wavelengths of the light, assuming that one beam is parallel to the Earth's motion and the other is at right angles. Such a shift could have been detected by Michelson and Morley; in fact they could easily have seen a shift of 0·02 wavelengths or $\frac{1}{20}$ of the expected effect. They found nothing!

This result, although unexpected, was consistent with several un-successful, but less sensitive, attempts to detect an aether wind dating at least as far back as 1810. In fact the failure of these earlier experiments had already provoked attempts, well before the date of the Michelson–Morley experiment, to explain the absence of an aether wind by arguing that the Earth drags the local aether along with it. However, this argument ran foul of the observations of stellar aberration, a phenomenon which was discovered by Bradley in 1726 (see page 99). Simple geometrical arguments show that these observations can only be explained if the aether is stationary in space and does not move with the Earth.

This failure to find an aether came to be seen in retrospect as a positive step forward. For one thing it verified one of the basic assumptions of the theory of relativity; for another, it emphasized that there are some features of the physical world which we cannot understand in terms of common experience. In the 17th century Newton had shown us how to cope with a similar mystery, but his example had not been followed. Although Newton could offer no physical explanation of gravity, about which he was personally unhappy, he nevertheless put forward a working theory of gravity which explained all the motions of the solar system. He showed how scientists, like everyone else, must live with a certain amount of mystery and learn to make the best of it.

And so, as in *The Hunting of the Snark*, they never found the aether; it turned out to be a *Boojum* instead, and there is no clear description of a *Boojum*. The original problem, a medium to support the waves of light, disappeared. The photon theory of light, which came later, no longer needed it; we now know that light behaves both as a wave *and* as a particle. The aether was no longer needed to transmit gravity as a mechanical force; gravity became an enigmatic property of space itself. In fact everything for which the aether was required in the 19th century was explained without the aid of anything mechanical. The space where the aether used to be is now a lumber room for mathematical symbols and, rather sadly, we are left without a satisfactory mental picture of what goes on. In place of mechanism the modern physicist has given us mathematics, which brings us to the next topic, relativity.

The theory of relativity

The theory of relativity, like Newton's laws of motion, is often presented as though it was a bolt from the blue, a work of art without provenance. Looking back, however, we can see that it was a logical, almost inevitable, step forward in our understanding of space and time and of the relationship of the Earth to the rest of the universe. As such the theory belongs as much to astronomy as to physics; furthermore, most of the tests of the theory have been astronomical because it is only outside the Earth that we can find the extreme conditions which they demand. To put the theory in perspective we must go back to Newton's laws of motion.

The first law tells us that a body continues in a state of rest or of uniform motion in a straight line unless it is acted upon by a force; the second tells us what effect a force has. We have seen that, in order to arrive at these laws, it was necessary to transcend our everyday experience and imagine a body moving through space unimpeded by earthly friction. But in this picture how can we be sure that we ourselves are not moving through space? And, if we are moving, are Newton's laws still true? The answer is to be found in the first law itself. It does not distinguish between rest and uniform motion, which is the same thing as saying that Newton's laws are true whether or

Newton investigating sunlight in a darkened room. His reflecting telescope (see p. 143) can be seen ▷ on the table. Engraving after J. A. Houston (1812–44).

not we are moving, provided we are moving at a uniform speed. When the laws refer to 'motion' they mean *relative motion*; we need not worry about whether or not we are moving with respect to some celestial scaffolding which is absolutely at rest.

Newton believed firmly in the existence of absolute space and absolute time. Absolute space, he wrote, exists 'without relation to anything external, it remains always similar and immovable; absolute time flows equably without relation to anything external'. Newton made it quite clear nevertheless that it is impossible to measure our movement with respect to absolute space by observing 'the positions of bodies in our regions'. In other words, because the laws of motion are the same for all observers who are moving uniformly, it is impossible to find out which observer is absolutely at rest. He underlined this point in the *Principia*, and, in so doing, made one of the first attempts to state a theory of relativity. He showed that his laws of motion are true for bodies in a given space, 'whether that space is at rest or moves uniformly in a straight line without any circular motion'. He went on to point out that all our measurements of position and motion are *relative*, and that it 'may well be that there is no body really at rest to which places and motions of others may be referred'.

This principle of relativity, so clearly expressed by Newton for the laws of mechanics, was not accepted in the 19th century as being true for the newly discovered laws of electromagnetism which govern the propagation of light. It was inconsistent with the established view that light is propagated as a wave-motion in a stationary aether. The surprising failure of the Michelson–Morley experiment in 1887 to find this aether was therefore a considerable skeleton in the cupboard of physics, until Einstein transformed the whole scene by extending the principle of relativity to the propagation of light.

Albert Einstein wrote his first paper on relativity in 1905 while working in the Patent Office in Berne, having failed to get an academic post at the Polytechnic in Zurich after his graduation in 1900. Guided by the firm conviction that it is fundamentally impossible to detect motion relative to absolute space, his first step, the Special Theory of Relativity, was to extend to all the laws of physics the principle of relativity which Newton had established for mechanics and uniform motion. In 1916, as a Professor of the Kaiser Wilhelm Institute in Berlin, he took the next step and published the General Theory of Relativity which is not restricted to uniform motion but applies to all observers no matter how they are moving.

The Special Theory of Relativity is based on the premise, taken by Einstein to be a fact of nature, that the speed of light is the *same for all observers* in uniform motion. In other words it makes no difference whether the source of light is moving towards or away from us; the light we receive will always appear to be travelling at the same speed. In the same way, it makes no difference if we ourselves are moving. This revolutionary idea

Albert Einstein by Jacob Epstein.

certainly explained the 19th century failure to detect any motion relative to a stationary aether, but it was completely contrary to the previous picture of how light travels. It implied that the whole common sense analogy between the passage of light through the aether and the motion of a boat on a fast flowing river on which all discussions of the Michelson-Morley experiment had been based, was complete nonsense.

In one bold stroke Einstein had thrown out the whole idea of a 'real' aether; more precisely he had removed its last mechanical property of motion. Translating his original words, 'the introduction of a luminiferous aether will prove superfluous, in as much as the view developed here will not require an ''absolute stationary space'' provided with special properties . . .'

If we follow Einstein's remarkable prescription and take the speed of light to be the same relative to all observers, it is comparatively simple to work out the consequences which, not surprisingly, are also remarkable. They are described in many popular books on relativity and are usually illustrated in terms of the most rapid means of transport in use at the time the book was written. Thus Newton illustrates his ideas on relativity by ships, Einstein by railway trains, later writers use aeroplanes, and the most recent school books talk about spaceships. If you read one of these books you will find that measurements which we make of fast-moving objects depend on how fast they are moving relative to us. For example, a moving object appears to contract in the direction in which it is moving; to be exact we see it as shortened by the factor $\sqrt{(1 - v^2/c^2)}$ where v is our relative speed and c is the velocity of light. Time is slowed down in the same way and by the same factor, and the same sort of thing happens to velocity, acceleration, mass, etc. The essential message is that all physical measurements are *relative*, not absolute.

At first sight this statement may look fairly humdrum, perhaps obvious, but that is deceptive. One of the great fascinations of science is that quite simple considerations often lead to conclusions which are profound and sometimes bizarre. This is certainly true of the Special Theory; let us consider two well-known examples.

The first deals with mass and energy. The simple equations of the Special Theory show us that the mass of a moving body appears to increase with speed; in fact it increases by the factor $(\sqrt{(1 - v^2/c^2)})^{-1}$. This formula can easily be rewritten to show that this increase in mass is exactly equal to the increase in kinetic energy of motion; we cannot distinguish between a gain in mass and a gain in energy. This result can be generalized to cover all forms of energy including, for example, heat, so that a gain in heat also behaves like a gain in mass, a conclusion faintly reminiscent of the old caloric theory! Einstein encapsulated this simple, but profound, conclusion in his famous formula, $E = mc^2$, his trade mark, which reminds us that mass (m) and energy (E) are equivalent. There is no need to underline the practical importance of this formula in the development of atomic energy or in our understanding of how energy is produced in the sun and stars.

The second example, 'the twin paradox', is bizarre. It has the meretricious quality of a conjuring trick, but in reverse; we can see exactly how the trick is done, but we don't believe that it really happens. The scenario is that twin brothers live on Earth and have identical clocks; one day one of the brothers sets off in a spaceship for a look at the nearest star, taking his clock with him. After a while he returns and compares his clock with the one at home and finds that they show different times. The traveller finds that he is younger than his stay-at-home brother, not because of a carefree life on the spaceship, but because the time-scale on his spaceship was slowed down relative to time on Earth. To be more precise, if the spaceship travels at about 87 per cent of the velocity of light, its time scale is slowed down to roughly half that on the Earth. Thus, if the twins were 20 years old at the start of the journey and the journey took 20 years by the clock on Earth, then, when they meet, the stay-at-home twin will be 40 years old and the travelling twin will be 30 years old; the travelling twin will have an extra 10 years to make up for all that boring time spent in the spaceship.

The story is called a paradox because, apparently, one ought to be able to look at the problem from the point of view of either of the twins and regard their motion as relative. Looked at in this way there seems to be no reason why one twin should be older than the other. The answer is to be found in the textbooks of relativity; it boils down to the argument that the situation is not symmetrical; one twin, the traveller, is in *accelerated* motion and therefore experiences a force which the other twin does not.

The importance of this 20th century parable is that it illustrates with disconcerting clarity the doctrine that time is relative, not absolute. And yet, the conclusion that one twin has *really* lived longer than the other is one that most of us cannot believe. Most of us, like the philosopher Henri Bergson, suspect that time in the spaceship only *looks as though it goes slowly*, in much the same way as people look smaller as they walk away. If by *really* we mean that it offends our common sense and is contrary to our experience of the everyday world then, obviously, we are right; but if we mean that we would be prepared to bet real money against Einstein in an actual experiment then, I believe we are wrong. Personally, I would put my money on the travelling twin being younger because, whenever it has been tested, the Special Theory of Relativity has been shown to be right. No one, I admit, has yet done the experiment with twins; it would be intensely interesting to see what odds the bookmakers would offer on the result.

The General Theory of Relativity is a far more difficult business than the Special Theory. It tells us that the laws of physics are the same for *all* observers, no matter how they are moving; it applies to accelerated motion and to gravitational fields, and is not restricted to observers moving uniformly in a straight line. When he published this theory in 1916 Einstein destroyed the very last traces of the ancient view that the universe has been arranged especially for us. The laws of physics, Einstein said, are the same for everyone, no matter where they are, and the whole idea of absolute time and absolute space is a misleading illusion from the past.

In developing this theory Einstein started from a simple fact about nature which has been known since the time of Galileo. The acceleration due to gravity is the same for bodies of different mass and different composition; in a vacuum a cannon-ball and a feather would fall together. If you push a cannon-ball and a feather and measure their accelerations you find their 'inertial masses'. If you weigh them and measure the pull of gravity you find their 'gravitational masses'. Experience shows that inertial mass is strictly proportional to gravitational mass for all objects; it is not understood why this should be so, but it certainly explains why all objects fall at the same speed. Einstein took it to be a fact of nature, just as he had previously taken it to be true that the velocity of light is the same for everyone. He went on to establish what is called the *principle of equivalence* of gravity and inertia which, in his popular works, he illustrated by what happens in a lift. Briefly, if the lift is falling freely, the Earth's gravity seems to disappear and a cannon-ball and a feather would float about in the air; a phenomenon to which television viewers are accustomed in programmes from spaceships. If the lift is accelerating upwards, gravity appears to increase, the cannon-ball and feather fall sharply to the floor, and we feel more weight on our feet. Einstein showed that over small regions of space the effects of acceleration and gravity are strictly indistinguishable, they are *equivalent*. This *principle of equivalence* was the missing link between Newton's laws of motion and his law of gravitation, and Einstein used it to formulate his celebrated field equations which tell us how space, mass, inertia, and gravitation are related.

Before we go any further we must say something about the idea of *space-time* which is central to the General Theory of Relativity. Most of us, I believe, are prepared to accept that space and time are mixed up in some way or other; every time we look at the sky we see stars that we know are not 'all there' at the same time. As I write, I can see the Southern Cross out of the window and I know that the light from the brightest star (Acrux) has taken about 300 years to reach me; it started at about the time that Newton published his *Principia*. On the other hand the light from the second brightest star (Mimosa) has taken about 450 years; it started about the time that Copernicus published his *De Revolutionibus*. But from elsewhere in the Galaxy things would be different, the relative positions of the two stars would change, as would the 'difference in time' between them. Clearly we need some sort of framework in which we can fix events in time and space and which will allow us to calculate how these events would appear to different observers. In the Theory of Relativity we are invited to picture this framework as an amalgam called space-time, a four-dimensional continuum in which time has been combined with three-dimensional space in such a way that the equations of relativity are still obeyed. For the Special Theory this was a fairly simple job, but for the General Theory it took years and involved mathematics which, regrettably, is too difficult for most physicists and astronomers to understand; in fact in scientific circles it is still quite respectable not to understand it.

The first thing we notice is that the General Theory of Relativity seems to turn physics into a branch of geometry, and very difficult geometry at that. Einstein's equations tell us that the 'structure' of space-time is determined by the objects in it; remove all the matter and energy and there is nothing left. This structure is *curved* by the presence of mass or energy and our path through space-time is therefore curved near a massive body and we experience what we call gravitation. The same thing is true of a light ray; near a massive body light does not travel in straight lines. Other strange things happen in strong gravitational fields. Time is slowed down, mass is increased, the lengths of objects are altered, and space itself becomes so curved that geometry no longer obeys the laws of Euclid.

All this sounds very odd indeed. The trouble is that the results of the General Theory have been translated out of their original language, mathematics, into words which we can understand. Unfortunately most of these words are themselves so deeply imbued with our concepts of space and time that they do not convey clearly the original sense of the theory. Perhaps if we had translated the theory into an older language, such as Latin or Greek, it would not have seemed so unfamiliar. After all, in the light of history, its ideas are not so startling. The concept of physics as a brand of geometry certainly inspired Pythagoras and Kepler. As for space, it is interesting to compare General Relativity with Plato's ideas. The space-time of relativity 'guides' material objects and exists by virtue of their presence; in Plato's world material objects exist by virtue of space ($\chi\omega\rho\alpha$) which 'nurses' them like a mother! Nevertheless, when all is said and done I suppose most of us find it hard to believe that our world is *really* like the world of General Relativity. Perhaps we should remember that the General Theory is no more than a description of the world, a mathematical metaphor, and the only sense in which we can say whether or not it is '*really* true' is to test its ability to describe and predict events which can be observed. If it gives the right answers then, for the time being, we may accept it as 'true'.

What evidence do we have that the Theory of Relativity can be trusted? In seeking the answer we must remember that it differs from the classical laws of Newton only when speed, gravity, or acceleration are abnormally high. Thus we should test the Special Theory by looking for the effects of very high speed and the General Theory by looking for the effects of very strong gravitation.

The obvious place to look for very high speeds is in a laboratory devoted to particle physics where protons, electrons, and other particles are accelerated almost to the speed of light. In such work the effects predicted by the Special Theory of Relativity are of such practical importance that they have been verified many times. As a first example the equivalence of mass and energy has been verified in a number of different nuclear reactions to a high degree of precision. As a second example, there are particles which have short lifetimes, after which they decay into other particles and into radiation. If these particles are accelerated to very high speeds then, as we have seen, the Special Theory predicts that their time-scale should be slowed down relative to laboratory time; thus, to the observer in the laboratory, their lifetimes should increase with speed. There is plenty of experimental evidence to show that this actually does happen, which leaves little doubt that the Special Theory of Relativity agrees with what is observed.

How can we be sure that the much more abstruse General Theory of Relativity is sound? As a start let us ask whether anyone, since Galileo, has checked that the principal assumption of the theory is correct. Are gravitational and inertial mass really proportional? Here we can get a clear answer; yes, they are. Baron Roland von Eötvös tested wood, water, asbestos, platinum, copper, and magnalium (manganese plus aluminium) in Budapest in 1922. He found that for all these materials the gravitational and inertial masses are in precisely the same proportion, any difference being less than 1 part in 100 million. More recently, in 1964 at Princeton, R. H. Dicke has improved the accuracy of this experiment, for aluminium and gold, by 100 times.

Now what about some of the predictions of the General Theory? Was Einstein right when he forecast that a ray of light passing close to the sun would be bent through an angle of 1·75 seconds of arc? As we might expect, astronomers were very concerned about this particular prediction, but they had to wait for the end of World War I before they could test it. In 1919 the Royal Society and the Royal Astronomical Society organized two expeditions, one to Brazil and one to West Africa, to observe a total eclipse of the sun. A total eclipse allows us to photograph stars near the sun and to measure the bending of light by comparing their positions with those measured when the sun is elsewhere in the sky. The West African expedition found that stars near the sun were displaced by 1·61 seconds of arc with a margin of error of 0·40 seconds; the expedition to Brazil measured 1·98 seconds with a margin of error of 0·16 seconds. These results agreed reasonably well with Einstein's prediction and were hailed as a great triumph, not only for Einstein and relativity, but also for international co-operation in science so soon after a war. A British expedition had verified a German theory.

In recent years more accurate ways of doing this experiment have been developed using radio waves which, in all respects, are the same as light waves, except that they are longer. One technique is to measure the angular separation between two powerful radio sources when the radio waves pass close to the sun. Using waves a few centimetres in length, such experiments have confirmed that radio waves are deflected near the sun by an angle which agrees closely (better than 10 per cent) with the General Theory. An even more precise method was first tried in 1966. In this technique a radar pulse is sent out from Earth and is reflected by a planet or is returned by a spacecraft, and its travel time is measured with very high precision. When the path of this pulse takes it close to the sun then, according to the theory, it becomes curved and the travel time is increased. Experiments have been made using pulses reflected from Venus and Mercury and also pulses returned from Mariner spacecraft. These tests have also confirmed the

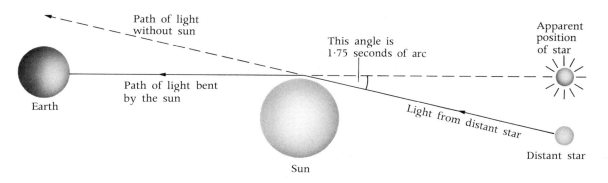

The Theory of Relativity tells us that when the light from a distant star travels close to the sun it is deflected through an angle of 1·75 seconds of arc. The apparent position of a star seen close to the sun is therefore shifted by that amount.

predictions of the General Theory.

Another famous test of the General Theory is concerned with the motion of the planet Mercury. Newton's law of gravitation tells us that a planet moves in an ellipse around the sun and, in the absence of other disturbances, it follows exactly the same path every time. By comparison, the General Theory of Relativity tells us that this is not true; in effect the mass of the sun distorts the space-time of the solar system so that successive orbits of a planet do not occur in precisely the same place. For this reason the whole orbit of a planet revolves slowly round the sun. This is best seen in the orbit of Mercury; theory predicts that the perihelion of this orbit— the point of closest approach to the sun—rotates by 43 seconds of arc per century, a phenomenon which is called 'the advance of the perihelion of Mercury'.

A comparison of this prediction with actual observations of Mercury is complicated by the fact that other planets, notably Venus, influence the motion of Mercury and calculations show that its perihelion ought to advance by 531 seconds of arc per century. However, observations going back as far as 1765 show that the perihelion of Mercury actually advances by 574 seconds of arc. Thus there is an 'unexplained' advance of 43 seconds of arc per century which agrees precisely with the prediction of the General Theory of Relativity.

There is one more classical test of the General Theory which we must note because of its importance to astronomy. The theory tells us that the wavelength or colour of a light wave is affected by gravity. This is only to be expected if we accept Einstein's equivalence of mass and energy. A light wave obviously has energy and therefore mass and so, like any other mass, is affected by gravity. If light of a certain colour (wavelength) is emitted in a place where the pull of gravity is very strong, say on the surface of a massive star, and then travels away from the star to a place where the

113

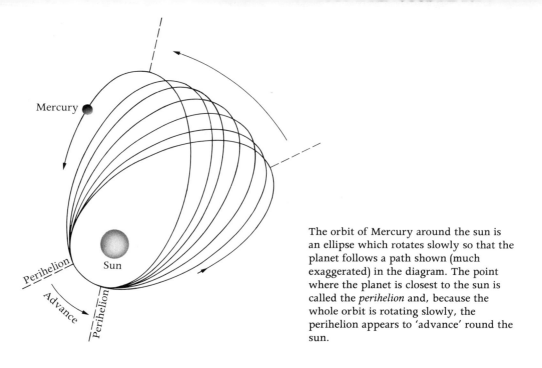

Mercury

Sun

Perihelion

Advance

Perihelion

The orbit of Mercury around the sun is an ellipse which rotates slowly so that the planet follows a path shown (much exaggerated) in the diagram. The point where the planet is closest to the sun is called the *perihelion* and, because the whole orbit is rotating slowly, the perihelion appears to 'advance' round the sun.

gravity is lower, say the Earth, theory tells us that it will be reddened, which means that its wavelength has been increased. This effect is known as the 'Einstein red-shift' and is obviously of importance to astronomy.

The nearest place to look for this red-shift is in the light from the sun; and there is a long and confused history of attempts to do so. The trouble is that the sun is not really massive enough to give a large red-shift. The expected change in wavelength in light from the sun is only about 2 parts in 1 million and attempts to measure it have run into a number of difficulties, one of which is the turbulence in the sun's atmosphere. Nevertheless there is at least one experiment which looks as though it really did verify the predicted red-shift. It was carried out in 1962 and measured the shift in wavelength of the characteristic light emitted by sodium vapour high in the sun's atmosphere.

A more promising place to look for the red-shift is on the surface of massive, compact, stars such as white dwarfs. Many attempts have been made to do this but it is very difficult, partly because the stars are so faint. In spite of this, observations have shown that white dwarfs do show a red-shift, but no really precise comparison has yet proved possible between this red-shift and theory.

The most recent evidence for the red-shift comes from physicists at Harvard. Using gamma-rays (which are like light waves but one million times shorter), they have succeeded in confirming the Einstein red-shift with an uncertainty of only 1 per cent!

All these tests confirm that the predictions made by the Theory of Relativity, both Special and General, are correct. In more technical language, the observations have so far failed to *falsify* the Theory. If, like absolute motion, we cannot hope to find absolute truth, then we conclude from these tests that the Theory of Relativity is nearer the truth than are Newton's laws. We must accept it as the best way of thinking about space and time until someone shows us the next step forward. There is plenty of scope for advance, particularly in uniting the laws of physics; in establishing, for example, the connection between gravity and inertia or between the laws of electromagnetism and the other laws. It looks as though there is enough for a future Newton or Einstein to do.

The principal tests of the Theory of Relativity are astronomical and are a valuable confirmation that the Theory is valid, and yet they fail to convey its importance to astronomy. The effects on which these tests are based are so small and difficult to measure, that one gets the impression that the Theory of Relativity is concerned with trivial corrections. However, this is far from true. The Theory involves a radical change in the way we think about space, time, and matter and is therefore of fundamental importance to astronomy. At the present time there are two fields in which this is highly relevant. The first is in understanding massive and compact stars, to which we shall refer later in this chapter; the second is in understanding the large scale structure of the universe, cosmology, to which we shall refer in the last chapter.

What happens to matter at very high temperatures: nuclear energy and the origin of the elements

There have been two particularly important advances in the relationship between astronomy and physics. The first, as we have just seen, was the publication of the *Principia* in 1687 in which Newton showed that the laws of physics, governing motion and gravity on Earth, also govern the bodies of the solar system. The second advance, which we shall come to now, was the discovery that the familiar chemical elements which we find on Earth are also present in the sun and stars. It was this discovery that finally destroyed the last traces of the ancient belief, that Heaven and Earth are made of different stuff.

In 1802, William Hyde Wollaston was using a prism to view sunlight entering a darkened room through a slit in the shutters of a room in London. Wollaston was very observant and he noticed that the spectrum of the sun is not a continuous band of colour but is crossed by several narrow dark lines; the light in certain colours is missing. He thought that these lines might perhaps mark the divisions between the primary colours but he had so much else to do, notably the refining of platinum, that he could not spare the time to investigate this odd phenomenon. A few years later, however, an optical instrument maker in Munich, Joseph Fraunhofer, also noticed the dark lines and used a new instrument, a spectroscope, to chart the position of 574 of them in the spectrum of the sun. The first explanation of these mysterious 'Fraunhofer lines' was given by a physicist, Gustav Kirchoff. Together with a chemist, Robert Bunsen—the inventor of the Bunsen burner—Kirchoff showed that, when light is passed through a gas, the elements and compounds in the gas absorb light at certain colours, producing dark lines in the spectrum. Furthermore, he established the crucial fact that each element or compound absorbs light at certain *characteristic* and *invariable* colours (wavelengths). Thus, by charting the lines due to various elements in the laboratory and comparing them with the lines in the solar spectrum, Kirchoff showed that certain elements are present in the sun; for example he identified no less than 60 lines in the solar spectrum due to iron, thereby demonstrating beyond reasonable doubt that there is iron in the sun.

The introduction of the spectroscope by Kirchoff revolutionized astronomy; it expanded a rather narrow preoccupation with celestial mechanics into a vigorous enquiry into the chemistry and physics of the stars. All that was known about the physical behaviour and chemical composition of matter on Earth could now be applied to the sun and stars, and the science of astrophysics came to life. A good example of the rapid progress is the classic paper published in 1929 by Henry Norris Russell in which he identified no less than 56 elements and 6 compounds in the sun and estimated their relative abundance. Since Russell's day spectroscopic analysis has been extended and we now have a fairly good idea of the chemical composition of most of the bodies, stars, comets, etc., which we can see in the sky. Nevertheless it still seems remarkable that we can find out what the sun is made of, without actually getting a spoonful to analyse. It is no wonder that the famous philosopher, Auguste Comte, who died only two years before Kirchoff published his discovery, wrote in his *Cours de Philosophie Positive* that to gain knowledge of the chemical composition of the stars is beyond our means.

Now that we have won this knowledge, thanks to the use of the spectroscope, have we learned anything really worthwhile about matter by looking up at the stars? The answer is emphatically yes, and the reason is that in the stars we can study the behaviour of matter at immensely high temperatures and pressures, far higher than anything we can now produce on Earth. It so happens that it is only under these extreme conditions that matter exhibits some of its most important properties. Let us look at one example of the behaviour of matter at very high temperatures which, as we shall see, is the key to one of the classical questions about the world—how did the elements originate?

Throughout history men have wondered how the diversity of matter which we find on Earth came into being. Was matter created in its infinite variety or did it evolve from something simpler and primeval? If we go back as far as we can, to about 600 B.C., we find that the most influential of the early Greek philosophers, Thales of Miletus, taught that all matter originates from one simple 'element', by processes which he did not describe; as his primeval 'element' he chose water. This basic idea has survived to the present day and has taken many forms. One of the best known was the idea put forward about 100 years after Thales by Heracleitos, who taught that everything originates from fire.

This idea was later elaborated into the notion that all earthly matter is compounded of four basic elements and the processes by which the mixing takes place were drawn from an analogy with living matter. It is obvious that a living thing transforms matter; a tree transforms air, water, and earth into wood. Why should this not be true of non-living things? No one was more interested in the answer to this question than the mediaeval alchemists who believed that all matter is descended from some primitive substance and that metals, like gold, breed in the ground. As Subtle says in Ben Johnson's play, *The Alchemist*, 'for 'twere absurd to think that nature in the earth bred gold perfect i' the instant; something went before. There must be remote matter'. Certainly some alchemists and astrologers thought that the growth of the seven metals took place under the influence of the seven major heavenly bodies, but no one realized until quite recently that this growth took place ages ago, not in the rocks, but in the stars.

The first real clue was found, as so often happens in science, in the course of trying to answer a completely different question—where does the sun get its energy? Attention was drawn to this problem in 1848 when Julius Mayer pointed out that the familiar processes of combustion, such as we experience on Earth, could not explain the prodigous output of energy from the sun. A simple calculation showed that if the sun was made of coal it would only shine for a few thousand years and, even in 1848, no one could believe that was long enough. The first credible answer was given in 1854 by the physicist Hermann von Helmholtz. He suggested that the sun is shrinking under the force of gravity and that therefore the gases of the solar atmosphere are compressed and heated. This idea was later examined by Lord Kelvin who showed that it leads to the conclusion that the sun is about 20 million years old, and that at the most it might last for 100 million years. When it was first put forward people were prepared to accept this explanation since there was no alternative. But, as the 19th century drew to a close, Lord Kelvin found himself fighting a losing battle against the geologists who were proving that the age of the Earth, and therefore of the sun, is much greater than 100 million years and probably exceeds 1000 million years. At the end of the century there seemed to be no way out of this impasse, until in 1905 Einstein threw the old idea that matter is indestructible out of the window and showed us that, in principle, it can be transformed into energy.

The way was now open to solve the problem of the sun's source of energy and it is interesting to see how clearly that great pioneer of astrophysics, Sir Arthur Eddington, put this problem to the nuclear physicists; in 1920 he spelled it out to them like an examination question. Eddington calculated the temperature and pressure at the centre of the sun and asked the physicists to discover a process which under these conditions would release the necessary amount of energy. He pointed out that it ought to be possible to build one atom of helium out of four atoms of hydrogen and that, if such a process could be found, it would release a very large amount of energy. This is because the mass of one helium atom is not, as one might expect, precisely 4 times the mass of one hydrogen atom, but slightly less, 3·97 times. Thus in the building of one helium atom 0·03 units of mass would be lost and must appear as energy. Many people, including Sir James Jeans, objected to this idea on the grounds that the centre of the sun would not be hot enough to bring about the necessary violent collisions between the atoms of hydrogen. Eddington stuck to his guns and told his critics to 'go and find a hotter place'.

It turned out that Eddington was right; the answer to his 'exam question' was given in 1938 by Hans Bethe, a physicist at Cornell University. We now know, largely due to Bethe's work, that there are two principal ways in which hydrogen is converted into helium inside a star. Let us look at one, the so-called proton–proton reaction. At the very high temperature, about 15 million degrees, inside the sun, the collisions between atoms are so violent that the nuclei of atoms are stripped of their electrons and hydrogen becomes a simple gas of protons and electrons. Nuclear theory shows that if two of these protons collide with sufficient violence they can 'fuse' together to form a heavier form of hydrogen nucleus, the deuteron. If now this deuteron collides violently with another proton it can 'fuse' with this proton to make a light form of the next element in the table of elements, helium (helium 3). Finally, two light helium nuclei can collide to form one nucleus of the common, heavier, form of helium (helium 4) plus two protons. The overall result of this chain of events is that four hydrogen nuclei are 'fused' into one helium nucleus and, in the process, slightly less than 1 per cent of the mass of the original hydrogen is released as energy. Although this doesn't sound very much it means that, by converting 10 per cent of its hydrogen into helium, the sun can go on shining with its present brightness for about 10 thousand million years, long enough to satisfy anyone.

The discovery of this process, the *fusion* of hydrogen into helium, solved the old problem of how the sun and the stars get their energy. But there is more to it than that, for it showed us where to look for the origin of the elements. We now believe that the original 'remote' matter was the simplest of all elements, hydrogen, and that all the other elements have been made, and are still being made, from primeval hydrogen. As we have seen, the first step is the conversion of hydrogen into helium which supplies the energy of stars for most of their lives. When stars grow old however, their internal temperatures rise and they start to draw their energy from the transmutation of heavier elements; for example, helium is transmuted into carbon, carbon into oxygen, oxygen into neon at successively higher and higher temperatures until iron is formed at a temperature of 3500 million degrees. At that point the star can gain no further energy because the reactions which produce elements heavier than iron absorb, rather than release energy. Thus for the origin of the heavier elements we must seek a different explanation and, at the present time most astrophysicists believe that they are produced in a number of minor processes, the most important being in supernovae. A supernova is a star which explodes with such fantastic violence that it blows most of its material out into space. There are good reasons to believe that in the hot envelope of a supernova the heavier elements are formed by collisions between neutrons and lighter elements.

Thus two important discoveries about how matter behaves at high temperature have been made by looking at the stars. We have found out how energy is produced by the transmutation of one element into another and we have learned where to look for the origin of the elements.

These discoveries have greatly enlarged our understanding of both physics and astronomy. Now that we know the source of stellar energy we can, quite literally, make sense of the otherwise bewildering variety of stars which we see in the sky. We can make mathematical models which tell us how the size and colour of a star vary throughout its life and we can estimate the age of the stars which we see. In much the same way the discovery that the heavier elements are synthesized in stars has made sense of the previously bewildering variety of the elements which we find on Earth. We now have some idea, albeit rather sketchy, of how and where these elements were formed and we can start to explain their relative abundance. The understanding of the part played by nuclear reactions in a star is, to my mind, one of the most satisfying achievements of modern science. It links stars to atomic nuclei, the largest to the smallest things we know; it has brought astronomy and physics together in a fruitful co-operation which has enlarged and invigorated both of them.

What happens to matter at very high pressures: white dwarfs, neutron stars, and black holes

White dwarfs

The first sign that there is anything queer about matter at very high pressure was the discovery of the Dark Companion of Sirius or, as astronomers call it, Sirius B. It is an unusually difficult star to observe, being so close to the much brighter Sirius A. As a result it was first detected, not by eye, but by measuring the very small changes which it causes in the position of Sirius A. This was done by Bessel in 1844 and Sirius B was, I suppose, the first 'invisible' star to be accepted by astronomers without actually being seen. In fact it was not seen until 1862 when an optician, Alvan Clark, noticed it by accident when testing the lens of an 18-inch telescope. If I may digress for a moment, there are many stories about Sirius and one which I find particularly intriguing is a report by two French anthropologists that the Dogon tribe of Africa maintain, as part of their religious and cultural traditions, that Sirius has a dark companion which is small, heavy, and invisible. It would be interesting to find out just how old that tradition really is.

The significance of the Companion of Sirius was not appreciated until 1914 when Walter Adams observed its spectrum and found that it has a surface temperature of 8000 degrees, much hotter than the sun. It was then possible to work out its size and mass and, to everyone's surprise, it turned out that Sirius B is incredibly small and dense. It has about the same mass as the sun and yet is fifty times smaller, no larger than the planet Uranus; it must therefore be about one hundred thousand times more dense than the sun and a piece about the size of a matchbox would weigh about 1 tonne. Other similar stars were observed and it was soon recognized that Sirius B is not an oddity but is a member of a new class of stars which astronomers call *white dwarfs*, because they are white-hot and small.

The most popular explanation of these white dwarfs has always been that they are very old stars. The argument is simple and goes like this. A normal star, like the sun, is a huge ball of hot gas which is in equilibrium between two major forces, the force of gravity trying to make it contract and the pressure of the hot gas inside it trying to make it expand. We now know that in a normal star these two forces are kept in balance by a self-regulating system. If the star starts to contract, then the temperature and pressure inside it rise and increase the yield of nuclear energy which, in turn, raises the gas pressure and opposes the original contraction; if the star starts to expand the reverse process takes place. The result is that, for most of its life, a star is perfectly stable and stays about the same size. But what happens when a star runs out of nuclear energy and this self-regulating system fails? Does it collapse like a pricked balloon?

Classical theory tells us that, when the energy supply fails, the star must cool down until its temperature falls to the temperature of space, close to absolute zero. It also tells us that the gas pressure inside the star will decrease and that gravity will force the star to contract. We should therefore expect to find a population of old stars which are small and dense.

So far, so good—but, as Eddington pointed out in 1926, classical theory does not tell us what will actually happen; it fails to explain how a star can ever reach the enormous densities of white dwarfs. It was certain that to

120

reach these densities the material of the star must be *ionized*; that is to say, the outer electrons must be stripped from their parent atoms so that the nuclei of the atoms can get closer together. And how, asked Eddington, is this ionized matter going to cool down? At low temperature it must solidify into normal un-ionized matter but, to do this, it will require energy which it doesn't have; it has radiated away its store of energy into space. In short, classical theory led to the paradox that, in order to cool down, a white dwarf requires energy which it doesn't have and can't get.

I get the impression from reading his books that Eddington enjoyed a good paradox; if so, his enjoyment was soon cut short by one of his colleagues at Cambridge. In 1926 R. H. Fowler published a paper on *Dense Matter* in which he applied some of the latest ideas of theoretical physics to the problem of the white dwarfs. Classical theory treats the molecules and atoms of a gas as completely independent particles which move about at random with velocities which increase with temperature. For such a gas it is simple to work out *gas laws* which relate the pressure, temperature, volume, and energy. What Fowler showed was that these laws cannot be applied to a gas compressed to the density of a white dwarf. The pressure is then almost entirely due to electrons which have been freed from their parent atoms and, at these extreme densities, we cannot treat these electrons as independent particles. They are then so crowded that they obey some of the rules which govern their behaviour inside atoms. The most important of these rules is called *Pauli's exclusion principle* and states that not more than one electron can occupy precisely the same 'state'. In other words electrons are no longer independent but share out the range of possible 'states' (e.g. momentum, etc.) between them in such a way that the classical gas laws no longer apply and the properties of an 'electron gas' are profoundly different from those of a conventional gas. The scientific term for matter in this compressed state is *degenerate*.

Fowler worked out the properties of degenerate matter and showed that they are just what is needed to explain the white dwarfs. The most striking feature is that the velocities of the electrons in the gas are governed solely by the density of the degenerate matter and, unlike an ordinary gas, are almost independent of the temperature. As a consequence of this remarkable fact degenerate matter exerts a high pressure *even when it is very cold*, which explains how a white dwarf can cool down to the lowest possible temperature without collapsing. Furthermore, the electrons retain their velocities, and therefore their energies, at these low temperatures, which explains how the material of a white dwarf can remain ionized although cold. It also explains Eddington's paradox; the degenerate material always retains sufficient energy, even when cold, to return to un-ionized material if the pressure is removed. In other respects degenerate matter reminds us more of molten metal than anything else; it conducts heat well and it stands high pressure without an appreciable change of volume.

Thus the discovery of the strange properties of degenerate matter solved Eddington's paradox and the problem of the white dwarfs. It was an important step forward in our understanding of the behaviour of matter at high densities, and it is interesting to see how such knowledge may be of practical importance. The densities reached in some of the new techniques of fusion, in which pellets are compressed by laser beams, approach those of the material in a white dwarf. To an astronomer the discovery of degenerate matter was of great importance because it showed one way in which a star can grow old without collapsing; it completed an important chapter in the life history of an evolving star.

Neutron stars

Is there really only one stable form in which a star can end its days? This question prompted astronomers to ask whether all old stars must become white dwarfs, and physicists to ask what would happen to matter at even higher densities. If, for example, we were to add more and more mass to the outside of a white dwarf would it eventually collapse? There is now little doubt that the answer to this question is yes; theory shows that when the mass of a white dwarf exceeds about $1\frac{1}{2}$ times the mass of the sun, the electrons at its centre would be crushed out of existence. They would combine with the protons in the atomic nuclei to form neutrons and the electron pressure at the centre of the star would begin to fail and the whole thing would collapse. We are not sure exactly what would happen next because some of the star may be blown away, nevertheless it seems fairly certain that at least some of the star would end up in another stable, incredibly dense, state. As the star collapses the density of the matter at its centre rises until it reaches about 10^{14} grams per cubic centimetre—roughly 1000 million tonnes in a matchbox—when most of the original electrons and protons will have combined to form neutrons. These neutrons, like electrons, obey Pauli's exclusion principle and form a degenerate 'neutron gas' with properties similar to the 'electron gas' in a white dwarf. They are, however, capable of supporting a much higher pressure at a much higher density and so the material of the star can now come to rest in a new stable state where the force of gravity is supported by neutrons. Such an object is called a *neutron star*, and, although it may have the same mass as the sun, it has a radius of only 10 km; it is unimaginably dense!

The theory of a neutron star was first put forward in the 1930s and for many years remained the sort of scientific curiosity which is mentioned hastily in the last chapter of books on astronomy. In due course it turned out that it was a solution waiting for a problem, and in 1967 the problem appeared, the pulsar. A pulsar is detected by a radio-telescope as a source of very brief pulses of radio waves which are received at precisely equal intervals of time, typically, one second. To cut a long story short, the only astronomical object which seems likely to produce such pulses is a very rapidly rotating star with an intense magnetic field. Since it is to be expected that neutron stars spin very fast and have enormous magnetic fields it is widely believed by astronomers that they are responsible for pulsars. In fact the existence of pulsars has been generally accepted as confirmation that neutron stars really do exist.

Black holes

As far as we know at the present day a star can exist in only two stable forms after it has exhausted its nuclear fuel, as a white dwarf or as a neutron star. No other stable forms have yet been discovered. But, as we have just seen, theory tells us that a white dwarf would collapse under its own weight if its mass is greater than about $1\frac{1}{2}$ times the mass of the sun; it can also be shown that about twice this limit applies to the mass of a neutron star. What, then, happens to more massive stars?

Until a few years ago it was thought that a massive star could settle down as a white dwarf or neutron star by shedding its surplus mass into space. It could do so slowly as it passed through the 'giant' stage of its life cycle when its atmosphere is distended and rather loosely attached, or it could blow off its outer layers in a series of minor explosions such as we see from novae. Alternatively, a massive star might settle down by blowing off most of its mass in one almighty explosion (supernova) which would compress the material at its centre into a neutron star. To support this last idea we can point to the remnants of an exploded star in the Crab nebula at the centre of which there is a pulsar.

But now there is another possibility, the *black hole*, the *enfant terrible* of modern astronomy and the darling of space fiction. The latest idea is that a massive star when it runs out of nuclear fuel, may simply collapse. As far as anyone can see there is nothing to stop it doing so provided, of course, that it is too massive to become a white dwarf or a neutron star. Once such a star has started to collapse, theory tells us that the matter literally drops into the centre and within a few minutes the whole thing disappears out of sight. It leaves behind it a black hole in space, the more massive the star the larger the hole. Stars of moderate mass make holes about 10 km across. Einstein's theory of general relativity tells us that inside the holes the geometry of space is so curved that light cannot escape and so from the outside we see nothing; if we try to illuminate the hole the light simply disappears down the hole and so would we, if we were to get too close. From the outside the hole is therefore perfectly black and the only trace of the original star which survives is its gravitational field, rather like the grin of the Cheshire cat in *Alice in Wonderland*.

However, the true Wonderland of modern theoretical physics is not the outside but the inside of a black hole. Inside we are invited to wonder what happens to the matter when it reaches the centre. Our guide is the General Theory of Relativity and, apart from telling us that the matter can't escape

from the hole, it has little to say. In technical jargon the equations develop a 'singularity' at the centre of the hole where they predict that the density of the matter becomes infinite. In plain words we don't know what happens to matter at these extreme densities and we need a new theory.

This complete failure of theory is regarded by some of our prominent physicists as 'the greatest crisis in physics of all time'. I suppose we should sympathize with them, for it must be exasperating to see all one's algebra suddenly disappear down a cosmic drain without any tools to fish it out. Nevertheless I think we may take this crisis calmly. There have been many crises in the history of science and most of them have been resolved either by a new idea, like quantum theory, or by a new observation, like the Michelson–Morley experiment. Furthermore, whenever we touch on infinities of time or space, we must be prepared to meet with mystery. In the meantime we can entertain ourselves by reading the conjectures of those theoretical physicists who would have us believe that the matter at the centre of a black hole emerges in a different universe! At least they are reasonably well insured against disproof; if we can't see into a black hole, we are going to have difficulty in deciding who is right about what happens inside. I can't help feeling that we are repeating, in modern terms, mediaeval speculations about how many angels can stand on the head of a pin.

What is needed now is some solid observational evidence for a black hole, and this is not proving easy to get because looking for a black hole is like looking for a black cat in a dark cellar. At the present time the most promising technique is to look for double stars in which one component is massive and invisible. The difficulty with this method is to be sure that the invisible component is not an unusually dim sort of star. Nevertheless, there are already a few likely candidates for the first black hole, the best, to date, being the double star Cygnus X-1 where there is an invisible component with a mass of about 8 times the mass of the sun. It may well be that this object will be accepted as the first black hole; if so, it will give us experimental confirmation that matter really does behave in an exotic way at extremely high densities.

The way ahead

So much of this chapter is necessarily historical that it may have given the impression that the great days of the influence of astronomy on physics are past. This is not so. For instance, if we look at one of the younger branches of astronomy, e.g. radio-astronomy, we see that there are problems about the recently discovered *quasars* which may well have a profound influence

◁ The Crab nebula—the remains of a star which exploded in A.D. 1054.

on physics; only time will tell. There is the completely unanswered question of where these quasars get the vast supplies of energy which they radiate into space. The equivalent problem in the sun led, as we have seen, to significant advances in physics and maybe the same thing will happen again. Secondly, there is the highly controversial problem of their red-shifts. A red-shift is a displacement of the wavelength of the spectrum of the quasar towards longer wavelengths, that is to say, towards the red end of the spectrum. In normal galaxies such a displacement is usually taken to mean that the galaxy is moving away from the observer. The well-known conclusion that the universe is expanding (see Chapter 5) is based on the observation that all the galaxies are moving away from us; the farther away they are, the faster they move and the greater their red-shift. The main difficulty in accepting this interpretation for the red-shifts of quasars is that they are surprisingly large and correspond to distances which some astronomers find hard to believe. The conventional interpretation of red-shifts has therefore been questioned and it may well be that out of this lively controversy some basic advance in physics will emerge.

Another topic recently opened up by radio-astronomers is the existence in space of quite complex molecules, mainly carbon compounds. The

126

Two vignettes from the title-page of Tycho Brahe's *Astronomiae Instauratae Mechanica* (1598). *Left* Suspiciendo–Despicio (by looking up—I look down) shows the author with a globe and a pair of compasses in his hand. *Right* Despiciendo–Suspicio (by looking down—I look up) shows him looking down at a chemical apparatus and with a snake, symbolizing medical science, wound round his arm. Tycho was making the point that to comprehend what lies at our feet we must look to the sky and vice versa.

problems of how these molecules were formed and how they can survive in the inimical environment of space is of considerable interest to physicists, chemists, and scientists with exotic names like galacto-chemists and exo-biologists.

Then again, what new problems will arise from the work which is going on at the present time in ultra-violet, X-ray and gamma-ray astronomy? What new branches of science will they call into being? The more the labels on the departments of our universities proliferate and change, the more they will illustrate the essential point of this chapter, that science advances on an endless frontier. It may seem strange that if we want to comprehend what lies at our feet we must look at the sky. Nevertheless, the history of physics and astronomy plainly shows that to understand a star we must look at an atom and to understand an atom we must look at a star.

5 Stars and Man

In recent years the influence of science on society has been a favourite subject for books, talks, and committees of inquiry, particularly since the end of World War II in which the use of radar, jet aircraft, penicillin, and operational research drew attention to the efficacy of science in solving practical problems. In these debates the importance and value of the applications of science have never seriously been in doubt, except perhaps by some of the more ardent exponents of the counter-culture; you have only to outline the benefits of research into health and wealth to make a convincing case. But the value of the whole of science has not always been accepted, especially of those branches which are removed by one or more steps from practical application and which are usually dignified by the faintly ludicrous title of pure sciences.

Two principal themes run through the modern criticisms of science. One of these is the so-called humanist critique which really is nothing more than the age-old battle between the arts and sciences echoed in *Two Cultures* by C.P. Snow. In a nutshell, the humanist tells us that science is amoral, devoid of values, and dehumanizes society because it is about things and not people; look, we are told, at the development of atomic weapons. The second main theme is the demand for *social relevance*; what, we are asked, is the use or moral justification of spending money on looking at the stars, when there are so many urgent social problems? Science, we are told, should develop a more active conscience.

The most common defence of science takes its practical value to be self-evident and aims to show that, in the long run, the continued flow of good things from the cornucopia of science and technology depends upon the progress of pure science which is, so to speak, the seed corn of the final crop. We are reminded that the advance of medicine, which we value without question, depends upon the discovery of X-rays which in turn depends upon the advance of physics. Admittedly, there will be some bad things, like atomic weapons, but we must learn to be more selective in what we choose to develop. This argument is sound as far as it goes; it would not have appealed to the ancient Greeks, but is effective in an industrial civilization dominated by economics. As usually presented, however, the connections between the more abstract sciences and practical benefits are not made clear; although activities such as astronomy and pure mathematics *are* vital to the whole structure of science, as we have seen in Chapter 4, to most people they appear to be useless. A second line of defence appropriate to a more limited audience, is that science, like music or painting, is worthwhile for its own sake. The pursuit of truth is like the pursuit of beauty—*ars pro gratia artis*. Science is reduced to an ornament of society.

In my view these arguments for science are too shallow; they do not answer the humanist critique and they fail to point out its fundamental importance to matters which, for want of a better word, I shall call cultural. I distrust this word because it has overtones of a narrow range of sophisticated activities like poetry-reading and chamber music. I shall use it here in a much broader sense to embrace our attitudes to the great questions about the world, its purpose, and origin, and our own place in the scheme of things. Although the answers to these questions belong properly to religion and philosophy, in any age they embody our vision of the universe—in other words, they reflect our current cosmology.

Dante's model of the mediaeval universe

Properly speaking, the story of cosmology starts with the first cave-man who wondered what lay beyond the horizon, and it continues with an account of how this horizon has widened and receded with the centuries. Let us take up this story at a convenient place, not too far away and not too long ago, say in Europe at about the time that Dante wrote the *Commedia* (A.D. 1300). Dante was a great scholar and his picture of the universe is an authoritative summary of ideas at an important time in our intellectual history, a time when the surviving remains of Greek thought had been rediscovered by European scholars.

Dante pictured the Earth as a motionless sphere at the centre of the universe. Only half of this Earth is habitable; to keep safely within the bounds of Christian doctrine, there is no major land mass in the southern hemisphere. St. Paul tells us (Rom. 10) that the preachers of the gospel 'went into all the Earth, and their words unto the ends of the Earth'. Since, at the time of Dante, there was no record of any preacher having visited the southern hemisphere, it was clearly unwise to assume that there was habitable land there. To those of us who live in Australia, as I do, I offer the consolation that Dante puts Purgatory in the southern hemisphere. Purgatory is a hill rising steeply out of the ocean and is precisely antipodean to Jerusalem; it is a staging post for reformed souls on their way to Heaven.

Circling around this stationary Earth there are nine transparent spheres. Eight of them carry the sun, moon, planets, and stars; the ninth, the primum mobile, drives the whole system of spheres like a merry-go-round. This ninth sphere is the boundary of space as we understand it. Outside it is

OVI COELVM CECINIT MEDIVMQVE IMVMQVE TRIBVNAL   LVSTRAVITQVE ANIMO CVNCTA POETA SVO   DOCTVS ADEST DANTES SVA QVEM FLORENTIA SAEPE
SENSIT CONSILIIS AC PIETATE PATREM   NIL POTVIT TANTO MORS SAEVA NOCERE POETAE   QVEM VIVVM VIRTVS CARMEN IMAGO FACIT

Dante and his Poem, a late fifteenth-century painting by Domenico di Michelino. On the left is the gate of hell and the sinners can be seen descending into the depths underground. Over Dante's right hand is the mountain of purgatory, divided into levels according to the sins which are being purged. On top of the mountain is the earthly paradise, or Garden of Eden. Above the earthly paradise are eight spheres of heaven, beginning with the moon and ending with the sphere of the fixed stars. Dante climbs up through the levels of purgatory and then proceeds through the various spheres of heaven until he reaches the Empyrean, the abode of God and the blessed.

Heaven, the abode of God, which cannot be described in terms of space and time. To ask what is outside Heaven has no meaning. The answer is certainly not a void; Aristotle had shown conclusively that a void could not exist. Hell, I should mention, is a conical cavity inside the Earth with Lucifer at the apex of the cone which marks the centre of the Earth.

The model is divided by the moon into two classes, rather like a railway train. The first-class compartments are above the moon and in them everything is perfect and unchanging and everything moves in perfect circles at a uniform speed. The second class is below the moon; everything there is subject to change and decay and, as we saw in Chapter 1, even the stuff of which it is made is not as good as in the first class.

Although Dante was concerned that his model should be true, as well as poetic, it does not do justice to the astronomical knowledge of his day. His model was not intended for astronomers; it is a highly simplified outline of the universe for popular consumption and, in making it, astronomy had to yield first place to theology. Nevertheless, the main features of the model, the stationary Earth and the concentric revolving spheres, were correct in terms of the astronomy of that time. They were based on the models developed by Aristotle and Ptolemy more than a thousand years before. These early models had reached such a remarkable degree of sophistication that Aristotle's model incorporated no less than 55 concentric spheres, while Ptolemy's model was even more advanced and difficult to understand.

The physics of the model are based directly on the teachings of Aristotle. Dante, like Aristotle, pictures the behaviour of physical objects in terms of living things; he sees the world as an organism. Objects *aspire* to reach the centre of the Earth because it is natural for them to do so; heavenly bodies are moved by the love of God and, seeking to imitate that which they love, they move in circles because circular motion is perfect motion. Although it may seem strange to us, this *organismic* approach to physics was conventional in Dante's time, the view of the world as a *mechanism* came later.

Taking a more general point of view, the main characteristics of Dante's model are that it is finite, complete, and hierarchical. The ideas of religion, astronomy, and physics are welded neatly into one coherent whole. Unless you know where to look, you can't see the cracks. The man who was largely responsible for putting it into its final form and for papering over the cracks was a theologian, not an astronomer. St Thomas Aquinas carried out most of the hard work while Dante was still a boy. St Thomas was deeply convinced that, although revelation is more important than reason and observation, they are all valid ways of arriving at truth. To prove his point, St Thomas set about the task of reconciling the best science of the day, mostly the works of Aristotle, with the teachings of the Christian Church. This was not a straightforward job because there were many points on which Aristotle disagreed with the Church; in fact the reading of his works was prohibited during the early part of the 13th century. It is unlikely, for example, that Aristotle would have agreed that the southern hemisphere is uninhabited on the strength of a quotation from St Paul. Nevertheless, after a good deal of friction with the authorities of the Church, St Thomas was successful; he made the last great synthesis of science and religion, the *Summa Theologica*, a cultural dinosaur. However, like the dinosaur, it lasted for centuries and set the pattern, not only for Dante's *Commedia*, but for a major school of Christian theology.

A view of hell: *The Last Judgement* by the fifteenth-century painter Jan Van Eyck.

The hierarchical structure of the model is very strongly developed. It is feudal society reflected in the sky. The executive branch of Heaven is modelled on the nobility. There are nine orders of angels, one for each concentric sphere. The seraphim, cherubim, and thrones look after the primum mobile, the firmament of stars, and the sphere of Saturn; below them the dominations, virtues, and powers look after Jupiter, Mars, and the sun; below them the princedoms, archangels, and angels look after Venus, Mercury, and the sphere of the moon. Messages and commands are transmitted from God to man down this chain and are delivered by an angel or, in special cases like the Annunciation, by an archangel.

Man in Dante's universe stands at the bottom of a ladder looking up to Heaven. The answers to all the great questions about life, its purpose, and its rewards, are known to the Church and are to be found higher up the ladder. The answers to all the questions about nature—what we would call science—were known to ancient authorities, like Aristotle, and are to be found in books; the great days of science lie in the past. The essential point about this cosmology is that it is basically a religious affair, a cathedral of the mind. Man is made in the image of God and is at the centre of the universe which is deeply concerned with his affairs; quite literally the whole thing revolves about him.

Copernicus removes the Earth from the centre of the universe

The great synthesis of science and religion produced by the theologians of the 13th century was too successful; it made science a prisoner of theology. In works such as the *Summa Theologica* of St Thomas Aquinas, the Church built into its doctrines certain specific scientific ideas which, inevitably, came to be regarded as established truths. In later years, when the Church was on the defensive, it was unwise to question them. 'It is philosophically impossible', wrote St Thomas in *De Unitate Intellectus*, 'for divine faith to profess what the reason must regard as false'. St Thomas was making a logical, not a polemic point; unfortunately he did not foresee the damage that this dangerous doctrine would do, not only to science but to the Church itself. As we know from the trial of Galileo in 1633, it has been used, from time to time, to keep the lid on scientific progress.

The trouble with trying to keep science a prisoner is that it goes stale rather quickly. Consider three of the principal scientific ideas on which the mediaeval universe rested, that the Earth does not move, that bodies move according to their nature, and that there is an essential difference between matter above and below the moon. When the Church adopted the geocentric model of the universe in the 13th century these three ideas were

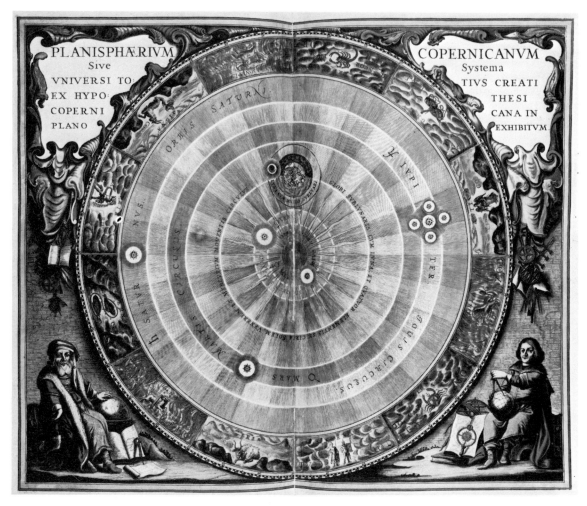

The Copernican Universe: an engraving from the Celestial Atlas, *Harmonia Macrocosmica*, of Andreas Cellarius, published in Amsterdam in 1660. The origin of the celestial maps by Cellarius is not known.

already going stale; they were vigorously questioned in the 14th century and by the end of the 17th century they had disappeared without trace. In short, the mediaeval cosmos was simply not true, and this fact was brought to light, as we shall see, mainly by astronomers.

The first idea, that the Earth is stationary, was challenged directly by a Canon of the Church, Nicolaus Copernicus, when he published *De Revolutionibus Orbium Coelestium* in 1543. Copernicus put the sun in its proper place very near the centre, and he made the Earth travel around the sun and rotate on its own axis. This idea had been put forward before, but had been put aside. In the 4th century B.C. Heracleides of Pontus taught that the Earth rotates on its axis and that the Heavens are stationary. About a century after him Aristarchos of Samos went even further and taught that the Earth rotates on its axis and travels around the sun. These models, however, were not popular and there was no obvious reason to accept them. For one thing, people preferred to think of man as being at the centre of the universe and, for another, their senses told them that the Earth is stationary. Looking at the stars on a calm night most people get the impression that the Earth is stationary and the sky is moving; they find it hard to believe that they are standing on a spinning Earth, moving sideways at more than 1000 km per hour. If you, too, feel the same way then you are in good company; even before the days of Aristotle and Ptolemy people raised objections to the idea of a spinning Earth. They argued that, if the Earth really spins, there ought to be a strong flow of air from the east accompanied, perhaps, by a mighty rushing sound; furthermore, an arrow shot vertically into the air should get left behind and fall to the west of the launching point. But these arguments failed to survive the vigorous criticisms which were made in the 14th century by such people as Nicole Oresme, Dean of Rouen, later Bishop of Lisieux. Although Oresme accepted the idea that the Earth is stationary, he refuted most of the standard objections to a spinning Earth in his interesting commentary on Aristotle, *Du Ciel et du Monde*. By the time Copernicus put this idea forward again, the way for it had been prepared.

De Revolutionibus made little impact on the man in the street; then, as now, he was more interested in astrology than astronomy. Many people, especially the clergy, would not have agreed that progress in cosmology is either possible or desirable. Martin Luther for example, was very scathing about Copernicus. Nevertheless, the learned world, both clerical and secular, gradually came to realize that *De Revolutionibus* was a serious challenge to orthodox cosmology and indeed, to the whole Aristotelian description of the world. Obviously it challenged the idea that the Earth does not move; furthermore, it raised doubts about two other beliefs which we noted as being important to the mediaeval cosmos. The belief that matter moves according to its 'nature' and that earthly matter therefore aspires to reach the centre of the Earth, is hard to reconcile with a model in which the Earth pursues a circular orbit around the sun. Again the belief that celestial matter (the fifth essence) is essentially different from matter on Earth is, to say the least, difficult to sustain in the face of a model in which the Earth behaves in the same way as the planets. Nevertheless the authorities of the Church didn't complain about *De Revolutionibus* so long as it was perfectly clear that the heliocentric model was a basis for computing celestial positions and nothing more. From the point of view of the Church the question of how to compute planetary positions was a minor technical detail. The major point was that God is concerned with man and has put him at the centre of the universe, admittedly a humble position in the mediaeval hierarchy, but nevertheless central. This truth was perfectly symbolized

135

by the mediaeval model of the cosmos, but not by a model in which the Earth moves and is not central. It was only when influential people, like Galileo, started to popularize the ideas of Copernicus and to insist that the Earth really does move, that the Church was forced to act. The earliest date at which we can be sure that *De Revolutionibus* was recognized as being revolutionary, in a cultural sense, is 1616; in that year it was put on the Index of Prohibited Books and was not removed until 1822.

Tycho Brahe charts the motions of the planets

Nearly 30 years after Copernicus published his book, mediaeval ideas about the universe were challenged again, this time by a new star and a bright comet. The new star was first seen in 1572, it outshone Venus, and disappeared only 16 months later. Fortunately, for the progress of astronomy it was seen by Tycho Brahe, one of the best observational astronomers in history. Tycho published his results in *De Stella Nova* in 1573. He also observed a very bright comet which appeared in 1577, and showed that it was at least six times farther away than the moon (*De Mundi Aetherii* 1588). Both these events were really serious challenges to orthodox

Tycho Brahe's great mural quadrant showing him and the various instruments he used at Uraniborg on the island of Hveen, Denmark.

ideas. The sudden appearance and disappearance of a bright new star was an obvious challenge to the well-established doctrine that the Heavens are perfect and changeless. The fact that the comet was above the moon, in a region which was thought to be changeless, and must have passed through the transparent concentric spheres of the planets was even more disturbing. If the solid spheres did not really exist, then the planets must be sailing through space without being attached to anything. What, it was asked, kept them in their orbits?

Tycho himself did not believe that Copernicus was right to put the sun in the centre of the universe. As a good Christian he acknowledged that a moving Earth was contrary to the teachings of the Church. As a good astronomer he asked why, if the Earth really does go around the sun, do we not see an annual displacement in the apparent positions of the stars? He looked for this displacement and couldn't find it; in fact it was not detected for another 250 years. To get round this problem Tycho invented his own cosmology in which the planets revolve around the sun and the sun revolves around a stationary and central Earth. This *Tychonic* system enjoyed a brief vogue in Europe but was superseded some years later by the work of Johannes Kepler. However, Tycho's main claim to fame, apart from the fact that he had a metal nose, is that he made a vast number of accurate observations of the planets from his magnificent observatory Uraniborg on the Danish island of Hveen. It was these data, the first really systematic and precise observations of the planets, which made it possible for Johannes Kepler to take the next decisive step towards a new model of the solar system.

Kepler shows us how the planets really move

All his life, Kepler sought for harmony in the universe and, at the age of 24, he thought he had found it. He conceived of a model of the solar system which reminds us of Plato who thought of everything in terms of geometry. Kepler put forward the idea that the orbits of the planets were designed by God to fit a concentric sequence of the five regular solids; a cube fitted between Saturn and Jupiter, a tetrahedron between Jupiter and Mars, a dodecahedron between Mars and Earth, and so on. This extraordinary model is of little interest now, but it did bring Kepler to work with Tycho Brahe. Tycho was an exceptionally difficult man to work with; after quarrelling with almost everybody who mattered, he left Denmark in 1597 and found a new patron in Bohemia, Emperor Rudolph II; in 1599 he moved to Prague where Kepler joined him in 1600. Tycho needed Kepler's mathematical ability to develop his own model of the universe (the Tychonic model); Kepler needed Tycho's observations of the planets to develop his own extraordinary model. In the event, Tycho's observations and Kepler's mathematics showed that both of their models were wrong.

Kepler, using Tycho's observations of Mars, showed that the movements of that planet could not be explained by any of the theories which had been put forward up to that time. He then proved, by the most laborious calculations, that Mars moves in an ellipse with the sun at one focus and he also showed that the speed of the planet in its orbit is not constant but varies in a calculable way. By 1609 he had made obsolete the time-honoured idea that all celestial bodies move in circles at a uniform speed.

Some ten years later Kepler discovered another important law of planetary motion which tells us that the square of the time which a planet takes to go around the sun is proportional to the cube of its distance from the sun. At last Kepler had discovered the sort of mathematical harmony which he had sought all his life.

Kepler's work is important to the development of cosmology not only because he replaced the old idea of perfect motion in a circle by a better idea, motion in an ellipse, but also because he started people thinking about what *causes* the planets to move. In the full title of his book *Astronomia Nova*, which tells us about his discoveries, he uses the word 'aetiological' which means 'concerned with causes'. In fact, orthodox cosmology already offered an explanation of why planets move; it was due to a moving spirit (anima motrix), but Kepler was not satisfied. He replaced the internal spirit (anima) by an external force (vis) which pushes the planet along, and he explained it as a magnetic force which emanates from the sun. In due course, this explanation proved to be wrong. Nevertheless, it was a step forward; in trying to explain the planetary system by a calculable mechanism Kepler was rescuing science from the old animistic ideas and was paving the way for the new scientific world-model of Newton.

Galileo shows us how to look at the sky with a telescope

Many of the new initiatives in astronomy (such as the development of radio-astronomy) are closely associated with the development of a novel instrument, usually something which gives us a new view of the universe. This was not true of the introduction of the Copernican model. The work of Copernicus, Tycho Brahe, and Kepler was done without a telescope and, until 1614, without logarithms. Astronomy's major tool, the telescope, did not arrive until 1609 when Galileo Galilei used it to look at the moon, the planets, and the stars. In his little book *The Starry Messenger* Galileo tells us what he saw. He saw mountains on the moon, more stars than can be seen with the naked eye and, wonderful to behold, satellites revolving around Jupiter. Later on, Galileo used his telescope to discover that Venus has crescent phases like the moon and that sometimes there are dark spots on the sun. None of these observations was accepted as conclusive proof that

Galileo Galilei,
1564–1642, from the
portrait by Sustermans.

the cosmology of Aristotle and Ptolemy was wrong, but they all pointed in that direction. In fact, Galileo used his telescope to draw attention to the shortcomings of the old system in a dramatic way which reached a much wider public than did the books of Kepler or Copernicus. The telescope gave people a new view of the sky, and what they saw persuaded many of them that the old system was wrong and that, maybe, there was something to be said for the new system of Copernicus. Galileo himself was an ardent Copernican and an opponent of the old cosmology. He was also not a man to let sleeping dogs lie. In his famous dialogue on *The Two Chief World Systems* (1632) he sets out clearly the relative merits of the Copernican (heliocentric) and the orthodox (geocentric) model of Aristotle and Ptolemy; furthermore, he wrote the book in Italian, not Latin, so that more people could read it. The Church had already warned Galileo not to promote the Copernican system, and so his book is thinly disguised as a discussion. In fact it is obvious propaganda for the idea that the sun is at the centre of the universe and that the Earth really does move around it. As one might expect, this was too much for the Church. The book was banned and Galileo was brought to trial in Rome in 1633. He was compelled to acknowledge that his ideas were wrong and was placed under house arrest for the rest of his life.

Two telescopes made and used by Galileo. The longer, 1·36 m in length, is of wood covered with paper. It has a biconvex lens of 26-mm aperture and 1·33-m focal length; a plano-concave eye-piece; and magnification ×14. The other, 0·92 m long, is of wood covered with leather with gold decorations. It has a biconvex objective of 16-mm useful aperture and 0·96-m focal length; a biconcave eye-piece (a later addition); and magnification ×20.

The silencing of Galileo is a classic example of locking the stable door after the horse has bolted. Copernicus, Tycho Brahe, and Kepler had already shown that the orthodox cosmology was wrong. What Galileo did was to confirm and publicize this fact by an entirely new means, the telescope. The time was ripe for a new model of the world and this time it was clear that it would have to be made by a scientist, not by a theologian.

Newton makes a new model of the world

The making of a new model from the ruins of the mediaeval cosmos took most of the 17th century. The French philosopher, René Descartes, made an impressive attempt but failed. The major problem was to choose the correct laws of motion and gravity and to show that they lead to Kepler's laws. Isaac Newton went a long way towards solving this problem when, at the age of 23, he left the University of Cambridge to work at home for two years (1665, 1666) to avoid the bubonic plague. We should, perhaps, never have known much about this part of his work if Edmond Halley, later Astronomer Royal, had not visited Newton many years afterwards in 1684. Halley and his friends at the Royal Society had tried to solve one of the most important problems in the astronomy of the time—why do the planets move in elliptical orbits? When Halley discussed this question with Newton he was astonished to find that Newton had already solved the problem; he hadn't told anyone about it and was not particularly anxious to do so. Fortunately, Halley was a persuasive, patient, and moderately well-off young man; his father was a successful soap-boiler. Halley persuaded Newton to publish his work and, to save Newton time and expense, he undertook to see the book through the printer and to pay any costs. As a result Newton's *Philosophiae Naturalis Principia Mathematica*, perhaps the greatest work in scientific literature, was published with an ecstatic dedicatory ode by Halley in 1687.

Part III of the *Principia* is called the *System of the World*. In it Newton showed that all the known motions of the Earth, moon, planets, and comets, even the rise and fall of the ocean tides, can be explained and calculated precisely in terms of his laws of motion and gravitation.

The Copernican revolution was now complete. The fact that the Earth goes around the sun was enshrined in a complete working system of the solar system which was greatly superior to anything that had gone before. It was superior not only because it predicted the positions of the planets more precisely, but also because it was scientifically more comprehensive; it was the first model to include both force and motion in mathematical form. Newton had achieved one of Kepler's ambitions; to borrow Kepler's phrase, he had founded a 'physics of the skies'.

The publication of the *Principia* had a profound effect on almost every branch of the culture of the day. Looking at the book itself one can't help wondering just how this came about. It is a formidable book, written in Latin and full of rather difficult geometrical propositions, and I often wonder how many people have read it right through. Nevertheless, within a century of its publication it went through 18 editions. To a large extent the immense influence of this book must have been due to the many popular versions which were written. Voltaire, for example, wrote a primer called *Les Eléments de la Philosophie de Newton*, presumably for the average man. Count Francesco Algarotti wrote a primer for the average woman, called *Il Newtonianismo per le Dame*.

Perhaps the *Principia* was such a success because the world was ready for a new model of the cosmos, something which embodied the spirit of the times. Lively discussions about science were taking place in the newly formed Royal Society (1662) and, as we saw in Chapter 3, there was a practical interest in improving astronomical theory for the purposes of navigating ships. Taking a broader view, the *Principia* stands at the end of a period of transition from the mediaeval world to the age of reason and enlightenment. The 18th century was a time of growing faith in reason and science, and the *Principia* reinforced that faith.

As one might expect from the controversy provoked by Galileo, Newton's new model of the universe had a disturbing effect on religious belief. From the Church's point of view—and we must remember that at this time in history orthodox religious belief was one of the most powerful

141

factors in the culture—there were major differences between Newton's *System of the World* and Dante's model of the universe. The principal difference was, of course, that in Newton's *System* the Earth is just one of many planets going around the sun and is no longer in the centre of the stage. Another major difference is that Newton's model is an exclusively scientific vision of the world, while Dante's is a synthesis of science and religion.

In Dante's cosmos the sphere of the moon divided Heaven from Earth. In Newton's cosmos this division disappeared and took Heaven with it; but it reappeared as a division in thought, between religion and science. Science had escaped from the prison of mediaeval religion and had set up an independent sphere, complete with its own sacred book, the *Principia*.

Newton, who was in fact more concerned about religion than science, was most anxious that his work should not challenge a belief in God. He pointed out that 'this most beautiful system of the sun, planets, and comets could only proceed from the counsel and dominion of an intelligent and powerful Being'. In other words, his model of the solar system is like a machine and a machine doesn't happen by accident, it must have a Maker. It is hard for us nowadays to feel the real force of this argument because we are accustomed to thinking about things as having evolved with time; for us the complex structure of the solar system or of an animal, can be explained in terms of gradual evolution from something simpler. But in Newton's day there was no such theory and people had to accept that complex and apparently purposeful things had been created from scratch; hence there must be a purposeful and rational Creator. Incidentally, it is often said that this belief in a rational Creator helped to pave the way for a basic tenet of science, the belief that the world itself is rational.

Newton's *System of the World* implied that there must be a Maker; but once He had set the whole thing going, there didn't seem to be anything for Him to do, although Newton thought that minor adjustments to the solar system would be needed from time to time. Nor was there anywhere obvious for Him to be. The idea of God became correspondingly more abstract. Newton tells us that God is 'always and everywhere, He constitutes duration and space'. He exists 'in a manner not at all corporeal, in a manner utterly unknown to us'. Man in this scientific universe could no longer think of himself as being made in the image of God.

This new image of God is nebulous compared with the robust and concrete picture painted by Dante. Nevertheless, it went down very well in the 18th century. Indeed it was seized upon by all those exponents of 'natural religion', like the Deists, who believed that all the things that really matter in religion could, like Newton's *System of the World*, be proved by pure reason. For them the Christian revelation and doctrines were either irrelevant or obstacles to true religion. We have only to read Voltaire, or David Hume's *Essay on Miracles*, to appreciate what forceful self-

The first reflecting telescope, made by Newton in 1668. The primary mirror has a radius of 2·5 cm and a focal length of 15 cm, giving a magnification of × 30. The external tube is about 23 cm long with a diameter of nearly 7 cm.

confidence this new rationalism inspired and how damaging it was to the authority of the Church.

However, in the long run it was not the rational arguments for the existence of God which had the greatest effect, it was the fact that the Earth had become just another planet of the sun. The geocentric model of the cosmos was the central idea in mediaeval thought from which religion, art, and literature drew their strength. The conclusion that this model was wrong, and that the Earth is just one of many planets revolving around a central sun, weakened the belief that Heaven is intimately concerned with our welfare and encouraged the suspicion that the supernatural world of the Church is little more than our own image reflected in the sky. As we shall see later the realization that the Earth is a planet of the sun was the first of several steps towards putting our place in the universe into its modern perspective.

Religion was not the only aspect of culture to be touched by Newton's work. His demonstration that the workings of the cosmos can be understood by rational enquiry encouraged would-be Newtons in almost every branch of thought. They tried to apply his technique of mathematical analysis of observations to an incredible variety of fields which included,

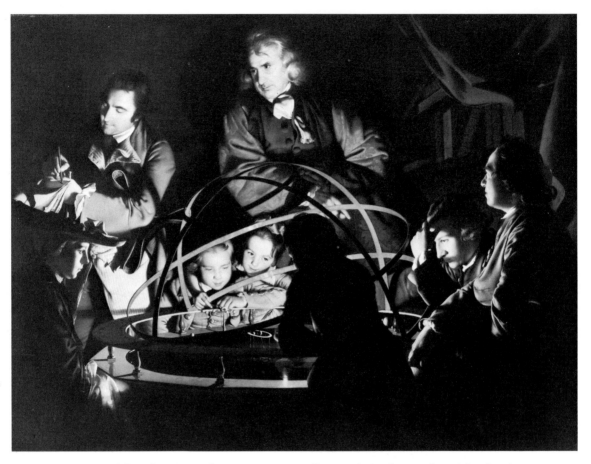

A philosopher giving a lecture on an orrery, from a painting by Joseph Wright of Derby. An orrery was a mechanical device for showing the relative motions of the planets by means of clockwork. The first one was made in the 18th century for Charles Boyle, Earl of Orrery, after whom it was named.

for example, political economy, moral philosophy, the social sciences, and so on. In many of these fields it was his technique, not his astronomy, which was the main influence and for that reason we will not discuss them here. There is, however, one more point which is worth mentioning. It is odd to find that Newton's *System of the World* is often described as a 'mechanical' or 'clockwork' universe. Usually we think of a machine as having an obvious prime mover, such as a motor or mainspring, which is connected to the moving parts by obvious mechanical linkages. But in Newton's model the planets revolve around the sun without any prime mover and there is no machinery to hold them in place. It is, in fact, quite unlike our idea of a machine. The planets, Newton tells us, move because there is nothing to stop them, and they are held in place by the invisible force of gravitation. He does not tell us how they came to be moving in the first place—in his view they were started by God; nor does he tell us how the colossal force

144

between the Earth and the sun reaches invisibly across space. This question is not trivial; a simple calculation shows that the force of gravity which pulls the Earth towards the sun is about 3 million million million tonnes. To transmit such a force we should need a steel cable about 7500 km in diameter and if we look into it a bit further, we find that it can't be done. Clearly Newton's model is not mechanical in the ordinary sense of the word. It is mechanical in the sense that it is repetitive and predictable; nothing living is involved and there are no angels to push the planets around the sun.

It is not surprising that Newton was criticized by some of his contemporaries, such as Leibnitz, for having introduced an 'occult' quality, gravitation, into science. It was said that his model depended on a miracle, action at a distance. Newton had nothing like that in mind; admittedly, he could not explain how the force of gravity worked, but he was convinced that in due course a *mechanical* explanation would be found. In a letter written in 1692 to the well-known scholar, Richard Bentley, he made it quite clear that the suggestion that a force can be transmitted through a vacuum without anything carrying it is 'so great an absurdity that I believe no man who has in philosophical matters a competent faculty of thinking, can ever fall into it'. As far as he was concerned a valid theory of the solar system demands that there should be a universal force of attraction between any two bodies. He would dearly have liked to know more about such a force, but it wasn't essential to his model; all he needed to know was how to represent it mathematically. It took nearly 200 years of trying to make mechanical models of Nature—of gravity, aether, light, etc.—to realize that it can't always be done. We no longer expect to understand gravity or light in the same mechanical sense that we understand a bicycle. In the meantime the whole idea that science is just a 'mechanical' explanation of the world has done extensive damage to our culture. It has encouraged the split between the sciences and the humanities. Only recently have people begun to realize that many of the basic concepts of modern science are just as abstract and imaginative as those of any other human activity.

In his *System of the World* Newton showed us how the solar system works and, in doing so, gave us an example of how to make progress in science. Science, he showed, is the art of the soluble; to make progress we must ask the right sort of questions, ones which we can hope to solve with answers that can be tested by observation. To understand how Nature works we must use reason, rather than revelation, and we must use observations to keep our feet on the ground. His work had a profound effect on all aspects of our culture and especially on religion; it altered the whole framework within which people thought about God.

Astronomy and astrology

One ancient way of thinking about Nature which was severely damaged by Newton's work was astrology. We know that at least 4000 years ago celestial omens were studied in Babylonia in order to determine the will of the gods in relation to public affairs. At that time there was a serious competitor, hepatoscopy, which was a technique for finding the will of the gods by inspecting the liver of a dead animal, usually a sheep which had been sacrificed. The theory of this method was that in accepting the sacrifice the God identified himself with the sheep, which then reflected his will in the structure of its liver, the liver being the seat of the soul. In its battle with hepatoscopy, astrology must have won the day. It was, I surmise, a more impressive and systematic technique and not so messy. From Babylonia, astrology spread to many countries and entered Greece somewhere about the 3rd century B.C. The Greeks put one of their own gods in charge of each planet, instead of the Babylonian gods, and developed an elaborate theory of the influence of the planets on the individual. This theory was based on the simple principle that events on Earth 'correspond' to events in the sky; in the words of astrology, 'as above, so below'. Thus a planet which is in the charge of Venus—Aphrodite in Greece—influences affairs of love—it controls beauty, elegance, love affairs, and so on. If a planet is in the charge of Mars—Ares in Greece—then its influence is warlike, forceful, and so on. All the qualities of the gods and goddesses on Mount Olympus are dispensed at birth to individuals on Earth through the medium of the planets, the mixture being determined by the exact time and date of birth, not conception. In modern jargon, astrology was 'personalized' by the Greeks.

Europe inherited Greek astrology much in the same way as it inherited Greek astronomy. Ptolemy is believed to have written the account of Greek astrology which is called the *Tetrabiblos*; like his book on astronomy (*The Almagest*), it became a standard text for European scholars in the Middle Ages.

Astrology did many of the things which science does today. It was a comprehensive theory of causes, an attempt to explain why things happen as they do. The proposition that the positions of the sun, moon, planets, and stars influence events on Earth must have started from the simple observation that the sun influences human affairs, and that different constellations of stars appear at different seasons. In due course this idea must have been extended to include the moon and planets, and so astrology began. Astrology provided a framework on which explanations of natural

Zodiacal man: an early fifteenth-century picture depicting the influence of the signs of the zodiac on ▷ the various parts of the human body. From the illuminated manuscript by the Limburg brothers, *Très Riches Heures*, produced for the Duke of Berry and now in the museum at Chantilly.

Aries. leo. sagittarius. sunt calida et sicca collerica masculina. Orientalia.

Taurus. virgo. capricornus sunt frigida et sicca melanco lica feminina. occidentalia.

Gemini. aquarius. libra. sunt calida et humida masculina sanguinea. occidentalia.

Cancer. scorpius. pisces. sunt frigida et humi da flemmatica feminina. Septentrionalia.

events and a system of making decisions could be built, especially where the facts were not known. In earlier times, before science gave us some of the answers, most natural events were inexplicable and unpredictable; for that reason, astrology grew to be relevant to almost everything that happened. Thus, in fields which we should now call science, it was important to understand the influence of the planets and stars on plants, minerals, chemicals, and so on. To take one example from medicine, the annual path of the sun among the stars (the zodiac) was associated with the human body projected on to a map of the sky. When treating the eyes of a man it was beneficial to remember that his right eye is under the influence of the sun, and that his left eye is under the influence of the moon; the reverse being true for a woman. In individual human affairs it was important to know precisely when people were born so that the planetary influences on their dispositions could be computed, and some of the major events in their lives foreseen. In public affairs it was important to compute the planetary influences when choosing a date for an important event. This, of course, required the services of a qualified astrologer and there were many official astrologers employed by the royal courts of Europe. In England the well-known astrologer Dr John Dee was invited by Lord Dudley to choose a propitious date for the coronation of Elizabeth I; judging from history he got it right. Dee, incidentally, was an able mathematician and a student of astronomy who was involved in the introduction of the Gregorian calendar into England, and yet most of his efforts, and certainly his income, were associated with astrology. He was frequently consulted by Elizabeth I and in everything except title he was the Astrologer Royal of that time. Had he been equally influential at court 100 years later, when the Royal Observatory at Greenwich was founded, he might well have been Astronomer Royal.

Lower down the social scale astrology reached the general public through the almanacs, which achieved a remarkably wide circulation after the invention of printing at the end of the 15th century. The almanacs carried the same sort of predictions and advice which we still see in our magazines and newspapers today.

The influence of astrology on all sections of the community was certainly great in the 15th and 16th centuries, and it would have been greater if it had not been opposed by the Church, and particularly by the Protestants. Ever since the days of St Augustine, the Catholic Church tolerated 'natural astrology' which is concerned with the influence of stars and planets on the natural world of minerals, plants, animals, and so on; in the eyes of the Church 'natural astrology' was just another branch of science. However, the Church did not approve of 'judicial astrology', which is based on the belief that the positions of the stars and planets at the moment of a person's birth, determine their character and, to a larger extent, their fate. As usually presented by astrologers this belief infringes on the domain of religion; it cannot be reconciled with Christian ideas about free will, the efficacy of prayer, or with the whole system of rewards and punishments symbolized by Heaven and Hell. The more diplomatic astrologers have insisted that a man's horoscope only helps him to know what sort of person he is, somewhat like psychoanalysis; the horoscope does not foretell an inescapable fate. The stars, they say, *incline*, they do not *compel* and, in that way, there is a loophole for free will and prayer. Despite these blandishments and the fact that some of the Popes had private astrologers, the Church has always been deeply suspicious of judicial astrology and, from time to time, has condemned it as, for example, in the Papal Bull of 1586.

In the long run, however, the real enemies of astrology were not the priests but the scientists, especially the astronomers. Essentially astrology is, like alchemy, a primitive theory based on how people, long ago, imagined the world might be. Inevitably it was replaced by science and so, when the French Academy of Sciences was founded in the 17th century, there was a ban placed on astrology. Thus, as our understanding of science grew from the 17th century onwards, astrology went the same way as Dante's universe because it was seen to be hopelessly old-fashioned; it was not disproved, it became obsolete. The same astronomical discoveries which had established Newton's *System of the World* helped to destroy the so-called 'rational' basis of astrology. The theory of planetary influence was weakened by the discovery that the Earth is itself a planet; the theory that comets are omens was demolished by the demonstration that they obey Newton's laws and that their appearances can be predicted. By calculating its orbit, Edmond Halley predicted that the great comet of 1682 would return in 1759; it came, but unfortunately Halley died in 1742, and wasn't there to see it.

By the beginning of the 18th century science had driven astrology from the mainstream of thought. It ceased to have any effect on the learned world and it lost its influence on public affairs. The last official astrologer in Europe retired in the 17th century. Nevertheless, astrology survived and flourishes today on a mixture of mild faith, curiosity, and entertainment. There are, we are told, more astrologers than astronomers practising today. It is easy to believe this to be true of Eastern countries, such as India, where astrology is still very active, but it is surprising to find that it was recently said of the U.S.A. Nowadays, scientists object to astrology because they see it as a magic practice masquerading as a science; in earlier days the Church used the reverse argument! Personally, I have never felt inclined to join these protests because I cannot believe that astrology does much real harm. In the past, it may even have done some good, when astrology and witchcraft offered alternative interpretations of Nature. A system which explained personal misfortune by an unfavourable conjunction of the planets was preferable to witchcraft which usually attributed it to the malevolence of a neighbour. Some people will always be attracted to a system of thought which interprets the world as being concerned with their personal affairs, which foretells the future, and helps them to make decisions. Many folk seek comfort and guidance and, nowadays, more and more of them seek it outside the Churches. To many people who consult astrologers, science is not an alternative; it cannot answer the sort of questions they ask, whether they should get married on a certain date, whether they will have children, whether they will be happy, rich, or live to a ripe old age. As long as astrology offers answers to such questions it will survive, but it has nothing significant to tell us about the nature of the universe.

Modern cosmology: space and time

From solar system to stars

The principal change in our picture of the cosmos since the days of Newton is that we have become aware of the immensity of the universe outside the solar system. You can see this from the textbooks on astronomy. A textbook published before the present century is mainly devoted to the solar system; at the most there is one chapter on the stars and a good deal of that is usually about the constellations. In comparison, the solar system usually takes up about one third of a modern textbook and the rest is about stars and galaxies. Let us look briefly at how this happened.

Both Dante and Newton were primarily concerned with the solar system. In Dante's model of the cosmos, the stars were little more than a decorative feature and were thought of as being all at the same distance. He looked down from the eighth cycle of Heaven—the circle of the fixed stars—and saw the rivers on the Earth a few thousand kilometres away. In Newton's *System of the World* we are not told much about the stars except that they are at immense distances from one another, otherwise they might 'fall on each other' due to gravitational attraction. How far away they are from us had never been established with any certainty. The only distance which was known accurately was the distance to the sun. It was measured in 1672 by observing the planet Mars simultaneously from France and French Guiana; the result, 138 million kilometres, is close to our modern value of 149·6 million kilometres. The first clue to the distance of the stars lay in recognizing that the sun itself is a star.

I am not sure exactly when this happened but the idea was certainly gaining ground in the 16th century. Given this basic fact, the distance to a star can be estimated by comparing some feature of it with the sun. One obvious feature which we can measure is how large the star looks in the sky. Assuming that the star is like the sun, then the ratio of its angular size (the angle subtended by the disc of the star at our eye) to that of the sun gives us its distance if we know the distance of the sun. Galileo used this simple method when he tried to answer a very serious criticism of the Copernican theory. It was the same question that had bothered Tycho Brahe—Why, if the Earth really goes round the sun, do the nearby stars not show a small annual change in their positions? Galileo suspected that the real reason was that the stars are too far away to show such a change. He decided to test this idea by finding out how far away the bright star Vega is from the Earth. To do this he hung a fine silk cord vertically and measured the distance at which he had to stand from this cord so that it just hid the star from his eye. From the distance and the width of the cord he calculated that the disc of the star subtends an angle of 5 seconds of arc and, by comparing this result with the angular size of the sun ($\frac{1}{2}°$), he showed that Vega is roughly 400 times further away than the sun. A simple calculation showed that, as the Earth travels its orbit, the change in the apparent direction of Vega would be too small to measure by any method known in those days. In this way Galileo refuted the opponents of Copernicus. Nowadays, we know that his argument was perfectly sound; it was sounder than he knew because he had overestimated the angular size of Vega. To measure the angular size of a star proved to be much more difficult than Galileo realized. The angular diameter of Vega was not known accurately until 1964 when it was measured at Narrabri Observatory (New South Wales); it was found to be 0·0032 seconds of arc, roughly 1500 times smaller than Galileo had found! Vega is very much further away than Galileo thought.

A simpler way of finding the distance of the stars is to measure the amount of light which they send us and compare it with the light we get from the sun. One of the first people to use this method was a contemporary of Newton, Christiaan Huygens, the inventor of the pendulum clock. He looked at the sun through a long tube with a tiny hole at the far end and adjusted the size of this hole until the sun looked just like the brightest star in the sky, Sirius. He reached the conclusion that the distance to Sirius was about 28 000 times greater than that of the sun. His result was far too small, but he was much nearer the truth than Galileo.

A third method—the one which produced the first really accurate result—takes us back again to the problem which worried Tycho Brahe and Galileo. As the Earth travels around the sun the apparent direction of the nearby stars, relative to the more distant stars, should show an annual change. In a period of six months the Earth moves, relative to the sun, by about 300 million kilometres; given the distance of a star, it is simple to calculate by how much its apparent position in the sky will appear to change. This change is called the *annual parallax* of the star and, even for the closest stars, is less than 1 second of arc. Such a small angle could not be measured until the first half of the 19th century when improvements in astronomical instruments made it possible. At once there was a race to measure the parallax of a nearby star. The winner is usually taken to be F. W. Bessel, an astronomer working at Königsberg. In 1838 Bessel measured the distance of the star 61 Cygni to be 660 000 times the distance of the sun or, very nearly, 100 million million kilometres. It was now clear that the universe of stars is vastly greater than the solar system and to measure this tremendous distance astronomers had to invent a new unit, the light-year, the distance travelled by a ray of light in one year. On this scale the distance to 61 Cygni is 10 light years, which sounds quite reasonable; the distance to the nearest star, Proxima Centauri, is merely 4·3 light-years.

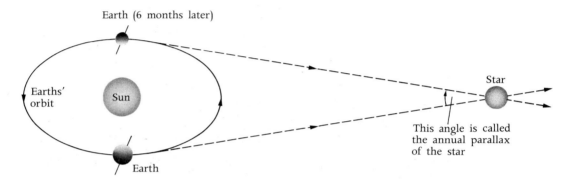

The direction in which a nearby star is seen from the Earth changes as the Earth goes around the sun. The maximum change is observed from two points 6 months apart and is called the annual parallax of the star.

From stars to Galaxy

In the meantime other astronomers were finding out what there was in all this vast space.

In Dante's cosmos the solar system was enclosed by crystal spheres rather like the walls of a mediaeval city. Many people speculated about what might lie outside these walls, but it was William Herschel who took the first really good look. He was able to do this because he made larger and better telescopes than anyone else.

Herschel was a professional musician who took up astronomy as a hobby at the age of 35. One evening in 1781 he was in the garden of his house in Bath looking at the sky through a telescope which he had made himself, when he saw what he thought was a comet. It turned out to be the planet Uranus; Herschel had discovered a new member of the solar system, the first to be discovered in recorded history. Shortly afterwards (1782) he gave up his career in music to devote the rest of his life to making telescopes and to looking at the sky; in his own words, he wanted to discover 'the construction of the Heavens'. Even a casual glance at the sky shows us that

A spiral galaxy in Coma Berenices. The galaxy is seen edge-on
as a flattened disc with a central bulge.

the stars are not distributed evenly; Herschel planned to map this
distribution, not only on a flat map of the night sky but also in depth. To do
this he counted the number of stars which he could see in different
directions with his telescopes and, from these counts, he computed how far
the system of stars extended. Together with his sister Caroline, one of
astronomy's saints, Herschel made an incredible number of these
observations, mostly from Slough. I have always been amazed that anyone
in Slough could have seen the sky for long enough to make all the
observations which William and Caroline undoubtedly did.

The result of all their hard work was to prove what many people had
already imagined. The solar system is a member of a vast system of stars,
millions of stars, which we now call the Galaxy. Herschel's work showed
that this Galaxy is shaped like a flat disc and that the sun is somewhere near
the centre. When we look at the Milky Way we are looking along this disc
and so we see a bright band of stars crossing the sky. He measured the
diameter of the disc to be about 800 times the distance of Sirius from the
sun, and its thickness to be one quarter of its diameter. Taking the modern
value for the distance of Sirius (8·7 light years) the diameter is about 7000

Reflecting telescope built by the Earl of Rosse. Construction was started in 1842 and it was ready in 1845, costing an estimated £20000. The mirror, weighing 4 tonnes, has a diameter of 6 feet (1·8 m) and a focal length of 54 feet (16·5 m). The telescope tube is 58 feet (17·7 m) long and 7 feet (2·1 m) in diameter.

light years. It is not surprising to find written on Herschel's grave, '*coelorum perrupit claustra*'—he broke through the barriers of Heaven.

Since Herschel's day we have learned a good deal more about the Galaxy and we now know that it is even larger than he thought. The modern picture is of a flattened disc of stars with a central bulge, rather like a fried egg. The diameter of the disc is about 100000 light years and the thickness is 400 light-years; within this disc there are about 100 billion (10^{11}) stars. The sun is not in the centre, but is about two-thirds of the way out from the centre to the edge of the disc. The disc itself has spiral arms and rotates about its centre once in 200 million years.

From Galaxy to universe

Among all the different types of stars which Herschel saw in the Galaxy there were some that didn't look quite right. When looked at with a large telescope they appeared fuzzy; he called them nebulae. He drew the attention of astronomers to the large number of these nebulous objects and suggested that they might be stars in the process of formation; alternatively, they might be other great systems of stars, the 'island universes' about which the philosopher Immanuel Kant had speculated some years before. However, no one had a powerful enough telescope to answer this question and Herschel's nebulae remained a mystery throughout the 19th century. A valiant attempt to get a better look at them was made by the Earl of Rosse in 1845. On his estate in Ireland, Birr Castle, the Earl built a 72-inch (1·6 m) telescope, a large telescope even by modern standards, and used it to look at these mysterious nebulae. He could see that some of them had spiral arms but he couldn't tell whether they were just cloudy objects or were made up of stars.

It was not long before someone built a telescope that would really do the job. The famous 100-inch (2·5 m) telescope at Mount Wilson (California) was completed in 1917. It is, by the way, hard to believe that this efficient piece of machinery was built only 70 years after the immense contraption of ropes and pulleys, wood, and brick walls, at Birr Castle. They differ, I suppose, by as much as the motor car differs from the horse and carriage. The 100-inch telescope succeeded where others had failed. Photographs taken of the two great nebulae in Andromeda and Triangulum showed that they were composed of stars or, at least, what looked like stars. Even so, it was still not certain what these nebulae were. Some people argued that they were small star systems inside our own Galaxy.

The 100-inch (2·5 m) reflecting telescope at Mount Wilson, California.

The argument was settled in 1924 by Edwin Hubble with the help of the 100-inch telescope. Hubble identified a specific type of star, a cepheid variable, in the two largest nebulae, in Andromeda and Triangulum. This particular type of star is the 'standard candle' of astronomy; by measuring its brightness, an astronomer can tell how far away it is. Hubble found that the distance to the nebula in Andromeda is 900 000 light years and the distance to the nebula in Triangulum is 720 000 light-years. It followed that the actual sizes of both these nebulae is roughly the same as our own Galaxy. Hubble had proved beyond any question that these two nebulae are distant 'island universes'; in modern terms they are galaxies.

Everywhere that Hubble looked with the 100-inch telescope he saw galaxies scattered through space. In his book, *The Realm of the Nebulae* (1936), he estimated that there must be about 100 million galaxies within the range of the 100-inch telescope and that the light from the faintest of them has travelled for over 1000 million years. The radio-astronomers now claim that they can 'see' even farther. They have discovered immensely powerful sources, *quasars*, which they believe to be the most distant objects which we can detect. Some of these quasars are believed to be so far away that the light and radio waves which they radiate has taken 10 000 million years to reach us!

This extraordinary increase in the size of our picture of the universe has taken place since the days of Newton. In brief, we now know that the sun is a common star in a Galaxy of millions of stars. This Galaxy is itself one of millions of other galaxies which stretch as far as our most powerful instruments can see. Our universe has become unimaginably large; but before we discuss that, let us look at some of the other changes in the picture.

The idea of evolution

As we have seen, the mediaeval picture of the world, Dante's model, was put together by joining Christianity to Greek science. The makers of that model adopted Greek ideas about circular motion in space, but they threw out the idea that time is also circular. A major school of Greek thought had said that the world is eternal and that it keeps on going round in cycles. The Christians, on the other hand, thought of the world as having a beginning and an end. The Book of Genesis tells us that, 'In the beginning God created Heaven and Earth' in one week. The references of the authorized version of the Bible tell us that all this happened in 4004 B.C. This rather mysterious date was computed in 1650 by Archbishop James Usher. He totted up the ages of all those unpronounceable old patriarchs in the Bible and, according to the 11th edition of the *Encyclopaedia Britannica*, he made two mistakes. The date should have been 4157 B.C.! Nowadays the whole idea that one can find out when the world began by ploughing through the Bible seems ludicrous, but not so long ago it was taken seriously, even by scientists. Sir

Isaac Newton thought that the date of Creation could be found in this way with an uncertainty, at the most, of 20 years.

However, the progress of science in the 18th century was particularly unfavourable to the Bible story of the Creation and Flood. The main attack came, not from astronomy but from what we now call the 'life sciences' and from geology. The sheer profusion of animals and plants which were discovered by explorers from the 16th century onwards, encouraged people to wonder whether these different animals and plants are related, and to doubt whether they could all have been created at the same time. One of the many problems in the story of the Flood is how the animals came to be distributed over the face of the Earth. People asked, for example, how the sloth got from Mount Ararat to South America. It couldn't have walked, so it must have been taken, but by whom? St Augustine worried about this too and suggested (*De Civitate Dei*) that the distribution of the animals after the Flood may have been expedited by a form of air transport—by angels. There were other problems. The time-scale demanded by geologists to explain their rocks and fossils was enormously longer than the Bible suggested. In the course of the 18th and 19th centuries it was gradually accepted that the Biblical account of the Creation and the Flood is not to be taken literally. Scientists have been trying to rewrite the Book of Genesis ever since and, as we shall see, they have still a long way to go. The first important step was to establish that all the complicated forms of life have evolved from something simpler. This was done in 1859 when the naturalist Charles Darwin published a book which is a major landmark in science and, indeed, in the history of thought. It was called *On the Origin of Species by means of Natural Selection or the Preservation of Favoured Races in the Struggle for Life*. As well as evoking the flavour of 19th century science, this title is almost self-explanatory. Darwin, and independently the naturalist Alfred Wallace, realized that the key to evolution is that organisms with favourable variations survive; the fittest survive. This deceptively simple idea brought order and explanation to the vast mass of observations about nature which were waiting to be explained. Darwin's theory of natural selection was as important to biology as Newton's theory of universal gravitation was to physics and astronomy.

Most people think of evolution in terms of biology; they have in mind man's descent from some awful creature swinging about in the trees. No doubt that is the most exciting part of evolution—people certainly thought so in the 19th century—but to an astronomer, biological evolution is only a small part of a long story. In the Book of Genesis only the last two days of Creation were devoted to making man and the animals. To rewrite the first four days we have got to find out how stars and planets, rivers, and rocks, came to be the way they are, and then we can talk about how the first traces of life appeared on Earth. We must find answers to the when, where, and what of the Creation of the Earth itself.

157

Modern cosmology : the Creation

Let us start at the beginning and find out what astronomy has to tell us about the Creation of the world. To do this we must consult a cosmologist because the days are long past when the current theory of the universe can be understood by practically anyone interested in science. Nowadays the tools of cosmology are the mathematics of general relativity and multi-dimensional geometry which are understood by only a few people. Our cosmologist would, I believe, tell us that the leading theory nowadays is the *Big Bang*.

The Big Bang

Like all modern theories the Big Bang was originally founded on one remarkable fact established by Hubble in 1929. Hubble observed the spectra of the light from distant galaxies with the 100-inch telescope at Mt Wilson. In the spectra there are prominent lines at certain wavelengths (colours) which are due to the absorption of light by specific elements such as hydrogen, helium, or calcium. The elementary physical theory of the Doppler effect tells us that if the galaxy is moving towards us, the wavelength of these lines is shifted towards shorter wavelengths, in other words the lines appear to be more blue than they really are. If the galaxy is moving away from us they are shifted towards longer wavelengths, they appear too red. Thus by measuring the wavelength of a particular line in the spectrum of a galaxy and by comparing it with the same line from a source of light in the laboratory, we can measure the velocity of the galaxy relative to the Earth. In 1929 Hubble showed that the lines in the spectrum of a distant galaxy are red-shifted and that this shift increases in direct proportion to the distance of the galaxy. If we accept the classical (Doppler) explanation of this red-shift—no other explanation is known at present—then the distant galaxies are moving away from us and, the farther away they are, the faster are they moving. In other words, Hubble showed that the universe is expanding.

Cosmology is necessarily a speculative branch of science with too many theories chasing too few facts. Prior to Hubble's work most theoretical models of the universe were based on the ancient, deep-rooted belief that the universe is both infinite in extent and unchanging in time.

These early models were unsuccessful; they all ran into technical difficulties of one sort or another—instability, infinite gravitational forces, etc. In retrospect one of these difficulties is particularly interesting. None of the models gave a satisfactory answer to the simple question—Why is the sky dark at night? At first sight the answer to this apparently rather silly question is that it is dark because the sun has set; but, if you think about it a

Edwin Hubble at the 48-inch (1·2 m) telescope at Mount Palomar, California. ▷

little more, there is a catch in it. If the universe is really infinite and unchanging then wherever we look in the sky our line of sight will eventually reach the surface of a star; the star may be very far away, but nevertheless, wherever we look there will be a star. The whole sky should therefore be covered in stars and should look as bright as the surface of the average star—in other words, as bright as the surface of the sun. Why doesn't it? This strange question was asked by Heinrich Olbers in 1826. It couldn't be answered satisfactorily in his day, but we now believe that we know the answer—the universe is not unchanging. It is expanding and simple theory tells us that the amount of light from a distant star is reduced if that star is travelling away from us at a speed which is comparable with the speed of light; furthermore stars have a finite life-time. These two effects, together, can explain why, in our expanding universe, the sky is so dark at night. It is startling to find that such a simple observation about the sky should have such an obscure, yet profound explanation.

When Hubble showed that the universe is expanding he gave cosmologists what they badly needed, a solid fact to work on. Now they knew what to look for, a model of the universe which expands and, at the same time, is consistent with modern ideas about space and time, that is to say, with general relativity. They hadn't far to look. Most of the groundwork had already been done by a Russian mathematician, Alexander Friedmann, in 1922.

Friedmann showed that there are two types of expanding universe. One type expands for ever; the other type expands up to a maximum size and then contracts. Which of these two types a model belongs to, depends upon how much matter it contains. If the amount of matter is below a certain critical limit, then the force of gravity is not strong enough to hold the universe together and it expands for ever; if the amount of matter is above this limit, then the force of gravity eventually stops the expansion and turns it into a contraction. A striking feature of these models is that, if we trace their history back to the start, we reach a moment when all the matter was condensed into a point of infinite density—like a black hole, but reversed in time.

The theory of the Big Bang is based on these simple expanding models. It has a disarming, primitive, simplicity. The universe, we are told, began with a violent explosion—a Big Bang—expanding away from a point where everything was infinitely hot and infinitely dense. The theory is modestly silent about the state of affairs prior to this explosion—if there ever was such a time—and immediately after it. Such questions, it says, lie outside our present knowledge of the laws of physics. If we choose to fill this silence with metaphysics, or with a reading from the first chapter of the Gospel of St John, we must be quick. About one thousandth of a second after the Big Bang the physicists claim to know what happened.

Their description of these first few seconds is still a bit confused. At the start the density and temperature was so high that no one is quite certain how matter behaved, but later on they are more confident. They tell us that the early universe was a fireball. It was so hot that nearly everything existed as radiation (gamma rays, X-rays, etc.) because the fundamental particles of matter (protons, electrons, neutrons, etc.) could not survive for long under those condtions. It seems that in one respect the Book of Genesis was right, *fiat lux*! As the primeval fireball expanded its temperature dropped and more and more of the radiation became matter, mostly hydrogen and helium. After about 1 million years, the temperature fell to about 5000 degrees (absolute) and there was then more matter than radiation in the universe. The stage was now set for the formation of the first material objects, the galaxies.

How galaxies and stars were made

The next part of the story is still rather vague. Astrophysicists believe that the galaxies condensed out of vast clouds of swirling matter, but just how they came to be the size they are, when they were formed, how they got their spiral arms, and so on, are still subjects of lively debate.

A cluster of distant galaxies in Coma Berenices. They are about 40 million light years away from Earth.

In the next stage of Creation the galaxies gave birth to the first generation of stars. Here we are on firmer ground. Indeed astrophysicists can give us quite a reasonable account of how the first stars formed from turbulent clouds of gas and dust in the galaxies, and of how later generations of stars are still being born today. Briefly, inside a dense cloud a large ball of gas and dust is pulled together by the force of gravity; the temperature in the centre of this ball rises until it gets so hot that nuclear energy is liberated and a star is born. The star then shines for most of its life by converting hydrogen into helium. When it gets old it converts helium into heavier and heavier elements until it reaches iron (see Chapter 4), at which point it runs out of energy and dies. If it is not much heavier than our sun, it can settle down to an almost infinite old age as a white dwarf. If it is much heavier, then it must lose weight either by blowing off its mass steadily, in puffs (novae), or in one almighty bang (supernovae). If it forms a supernova then most of the original mass is returned to the space between the stars, but a small dense remnant may be left (see Chapter 4) as a neutron star or, perhaps, as a black hole. The lifetime of a star depends mostly on its mass. A very large hot star lives a short but brilliant life of about 10 million years. An average, rather dull star like the sun has a lifetime of about 10000 million years.

How the Earth was made

So much for the Creation of the galaxies and the stars, but what about the Earth? How the Earth was formed is, of course, one of the great questions of science and the answer should not be rushed: it may take a very long time to be sure. The trouble is that we know so much about the Earth and the solar system and our knowledge is increasing so fast, that any theory has a hard time keeping up with the facts. The leading theory 50 years ago was very different from the theory today. At that time it was thought that the planets were formed from the sun in a near-collision with a passing star; the gravitational force of the passing star pulled out some of the surface of the sun and this stuff formed the planets. This idea had a powerful attraction for some people; it supported the belief that our solar system is unique and made it less likely that there are other inhabited worlds. It was welcomed by those who believed that man is unique and is the principal object of the whole of Creation.

However, this 'collision' theory was shown to be dynamically unsound and it was soon replaced by a much better idea which dates, at least in outline, from the 18th century. Broadly speaking, all modern theories are based on this old idea, that the sun and the solar system were both formed from one original nebula, the 'solar nebula'. The general idea is that, as the sun condensed from this nebula, it was spinning, as are most stars, and the centrifugal force of its rotation flung some of the hot material at its equator out of the sun into a thin disc. Somehow or other the planets were formed from the material in this disc. There are many difficulties in this model and each theory has its own way of surmounting them. For example, it has to be explained why most of the rotation in the solar system, more precisely the angular momentum, is in the planets and not in the sun. There are also problems in explaining the different chemical compositions of the planets. Nevertheless, there is general agreement that the sun and all the planets, including the Earth, were formed from the same nebula at the same time; furthermore it is now generally believed by astronomers that this process must be fairly common in the formation of stars. In short there must be other planetary systems like our own.

Why should we believe this story of Creation?

Before I try to answer that question, I must say that I take all theories about the origin of the universe, from the Book of Genesis onwards, with a large pinch of salt. In the absence of an eye-witness we are speculating about what happened 10 000 million years ago and it would not be surprising if we were wrong. Even so, we must do the best we can, so let us look at the evidence for the main feature of the story, the Big Bang. First of all there is the dramatic fact that all distant galaxies are travelling away from us; that, in itself, suggests an initial explosion. Secondly, it has recently been discovered that the whole sky is flooded with a faint glow which looks the same in all directions. The spectrum of this glow suggests that its source is at a very low temperature, about 3 degrees absolute, which is just what the protagonists of the Big Bang theory predicted. They claim that this glow is all that we can now see of the radiation in the primeval fireball. So far, no one has thought of a better explanation. Thirdly, the proportion of helium everywhere in the Galaxy is remarkably constant at roughly one quarter of the mass. This is just the amount of helium which, according to theory, is synthesized in the Big Bang. Finally, our estimates of the age of the oldest stars and of the rocks on Earth suggest that the age of the universe must be at least 10 billion (10^{10}) years. If we calculate, from the speed and distance of the receding galaxies, how long ago the Big Bang took place, we get much the same answer, 10 to 20 billion years (10^{10} to 2×10^{10} years). Until something else turns up, these are all good reasons for accepting that the Big Bang is the best theory we have.

Another idea: continuous creation

Before we leave the Big Bang it is worth remarking that it didn't reach its present position without competition. Its most serious rival for many years was the theory of *continuous creation*. This theory, also known as the *steady state*, is based on the idea that the universe did not begin with a bang, but has always looked much the same and will always continue to do so. The universe expands, but the density of matter, on a large scale, is kept roughly constant by the continuous creation of matter. Matter keeps popping up everywhere at the right rate to keep things looking much the same.

Many people have been attracted to this theory because they prefer the idea of a steady, eternal, universe which did not begin with a bang and will not disappear 'into thin air'. However, the progress of science does not respect preferences and the theory of continuous creation is currently in trouble. One objection which has been raised by the radio-astronomers is that, judging from what they see at great distances, the density of radio sources in space appears to have been much higher in the distant past than now. This result contradicts the idea that, on a large scale, the universe has not changed with time. Another objection is that the theory cannot account for the faint glow which, as we have just seen, is attributed to the radiation from the Big Bang. Personally, I would prefer to see the steady state theory in a healthier state; the competition between it and the Big Bang has made cosmology into quite a lively subject. If the theory dies, we shall miss it.

How did life begin?

Our cosmologist has told us how he would rewrite the first four days of Creation in the Book of Genesis. For a revised account of the last two days we must turn to the biologist. He will tell us that man and all other living things have evolved, step by step, from the first self-replicating molecule in the primeval slime; in other words life has evolved from the first molecule which, acting as a template, could produce other molecules like itself. But the biologist can't tell us yet how this happened or how long it took, and until he succeeds the two stories of evolution, the cosmological and the biological, cannot be told as one continuous drama from the Big Bang to man. They must, for the time being, be presented as two separate acts with an intermission.

To link these two stories it must be shown that the basic components of life, self-replicating protein molecules, can be produced in plausible circumstances from the basic elements of carbon, hydrogen, oxygen, and nitrogen. Again, this is one of those great questions of science, like the formation of the solar system, which may take a very long time to answer with confidence. The first major step towards this answer was the invention of the compound microscope early in the 17th century. The microscope has helped us to understand the nature of living things, just as the telescope has helped us to understand the stars. The discovery of spermatozoa by the 17th century microscopist, Anthony van Leeuwenhoek, and his work on insects helped to explain how living things reproduce themselves. His work on the reproduction of fleas, for example, dispelled the widespread idea that fleas are spontaneously generated by dirt. Louis Pasteur did the same thing for bacteria later on. I am reminded that in a recent lecture in Sydney, Sir George Porter remarked that the famous 17th century physician, van Helmont, gave a recipe for producing mice by spontaneous generation. A shirt, dirty with human sweat, is left in contact with wheat kernels for 21 days!

A more recent step forward in what is now called pre-biotic chemistry was taken in 1953 by Stanley Miller and Harold Urey at the University of Chicago. They passed an electric discharge through a mixture of water vapour, ammonia, and methane in a bottle. The conditions in the bottle were intended to simulate a thunderstorm in the primitive atmosphere of the Earth. When, after several days, they analysed the stuff in this bottle they found that it had taken the first step towards living matter; somehow or other amino acids—one of the building blocks from which all the proteins in living matter are built—had been produced. Perhaps that is the way in which living matter started on Earth.

There is another possibility, the first complex molecules needed to start life may have come from outside the Earth, not in flying saucers, but in the original material from which the Earth was formed. Radio-astronomers have detected a number of quite complicated molecules in the great clouds of gas and dust in interstellar space. Many of these molecules are compounds of carbon and are possible forerunners of the more complex molecules of living matter. How these molecules are formed in space is not yet understood but they are undoubtedly there in vast quantities. It has been suggested that they were present in the nebula from which the solar system was formed and initiated the development of life on Earth. Yet another possibility is that these 'pre-biotic' molecules came to Earth in meteorites, the stony debris which circulate in the solar system. There is a type of meteorite which contains carbon compounds and chemists have found complex molecules, even some of the amino acids, in several of these. In short, we don't know yet how living matter originated; we are not even sure whether it started from scratch on Earth or was helped from outside.

Are there other inhabited worlds?

Another important feature of our modern cosmology is that it makes the existence of other inhabited worlds not only credible but, some would argue, virtually certain. Two things have helped to bring this about. We have discovered that in our own Galaxy there are about 100 billion (10^{11}) stars, and that there are hundreds of millions of other galaxies within range of our telescopes. Also we now understand, at least in outline, how the solar system was formed together with the sun, and we can see no apparent reason why the same thing should not have happened around other similar stars. It seems possible that, in our own Galaxy alone, there may be as many as a billion (10^9) planets rather like the Earth. To say the least, it appears unlikely that our Earth is the only one to support life.

Space travel to the nearest star—*pace* science fiction—is well beyond our present capabilities and likely to remain so for as long as we can foresee. There seem to be only two ways of finding out about these other worlds; we could put out a 'welcome' mat for flying saucers or we could try to get in touch with them by radio. I don't know whether anyone has tried the first method, but they have certainly tried the second. In 1960 radio-astronomers used the 85-foot (26 m) radio telescope at the National Radio Astronomy Observatory in West Virginia (U.S.A.) to search for signals from the two nearest stars which are of the same type as the sun (Tau Ceti and Epsilon Eridani). They chose to listen at a wavelength of 21 cm, the wavelength radiated by the neutral hydrogen atom, on the argument that it is one of the fundamental wavelengths of nature and therefore likely to be chosen by one 'technically advanced' civilization to get in touch with another. I have always found this choice slightly baffling because, on Earth, transmissions at 21 cm are prohibited internationally to prevent interference with radio-astronomy. However that may be, they failed to get any signals.

If you look into this problem of radio signals in detail, it turns out that any civilization which wishes to get in touch with its neighbours must devote a really substantial effort to doing so. We cannot expect to pick up T.V. or radio programmes from another world, they would be far too weak. In the case of the two nearby stars (Tau Ceti and Epsilon Eridani) a few simple sums show that they could have been received in West Virginia only if they had used a very large radio-telescope (600 ft or 180 m in diameter) and a very powerful transmitter (1 million watts); furthermore, the transmissions would necessarily have been confined to a narrow beam and so this huge radio-telescope would have to have been pointed at the Earth. It is really not surprising that nothing was heard from those two stars. Quite apart from the size and power of the equipment needed to send signals, their civilization might already have been destroyed by one of the many dooms predicted for ours, or they might be a society of highly intelligent dolphins splashing happily about in the sea. All the same, if there are astronomers prepared to do it, and with money to spare, it does seem worthwhile to keep an ear open for our neighbours in space.

The recent acceptance of the idea that there are, very probably, other inhabited worlds is an important step towards seeing our Earth in the new perspective offered by modern cosmology. We shall return to that later.

166

'Miss! Oh, Miss! For God's sake, stop!'

Drawing by Whitney Darrow, Jr; © 1957 The New Yorker Magazine, Inc.

How will the world end?

Apart from telling us that we are as dust and can look forward to returning to dust, the Book of Genesis is mercifully silent about the future. However, later on in the Bible we are told that the world will end and that we shall all be judged. No specific timetable is given for this alarming programme, but the Book of Revelation leaves us with the impression that it won't be long. Modern cosmology has a lot to say about the past, but what has it got to say about the future? About the future of the Earth and of the whole solar system it is fairly definite. As the sun evolves it will become larger and hotter. At present this process is slow; it will take 2000 million years for the size and energy output of the sun to increase by a few per cent. In 5000 million years it will grow too hot and large for us to live on Earth and we shall have to find somewhere else to go. The astronomical version of the end of the world is egalitarian; the Earth will become a Hell for everyone, whether they have lived a blameless life or not.

About the future of the universe itself the cosmologists have little to say. They don't yet know whether our universe is expanding for ever, or expanding and then contracting. In principle this question can be settled in a number of different ways. The simplest method is to find the average density of matter in the universe by adding up the mass of all the galaxies. This density can then be compared with the critical density which divides the two kinds of universe. If we do this we find that the density of the visible universe adds up to only about 5 per cent of the critical value, which suggests that the universe is expanding for ever. This result, by itself, is not conclusive because we are not sure that we have taken into account all the matter in the universe; some of it may be in forms (e.g. black holes) which we cannot detect. There are however, two other methods. One is to measure the amount of deuterium (heavy hydrogen) in the Galaxy, and the other is to compare the age of the universe as estimated by the ages of stars and rocks with the age estimated from the recession of the galaxies. Both these methods are too complicated to discuss here, but they both point to the conclusion that the universe is expanding for ever. In due course our Galaxy with some of its close neighbours, will be left on its own. More and more of the matter will be locked up in white dwarfs, neutron stars, and black holes, and the Galaxy will itself become a stellar graveyard. But long before that happens, 5000 million years from now, the sun will have destroyed the Earth.

Modern cosmology and our world view

At first sight the most striking feature of our modern picture of the universe is its size; the distances have become unimaginably large. Popular books on astronomy usually make a great song and dance about this, but I find it hard to believe that any of the recent increases in size have had much effect on our world view. Once a distance, or a time, becomes unimaginably large, it makes little difference how many noughts we add. For most people, the distance to the sun is already unimaginable, and that has been known for centuries. Astonishment at the sheer size of the universe is a thing of the past. Thus, if we go back to the 17th century, we find the famous philosopher and mathematician, Blaise Pascal, complaining that all the empty space outside the Earth frightens him—*'le silence éternel de ces espaces m'effraie'*. And yet, in our own time people have become quite blasé. The modern poet Robert Frost writes:

> *They cannot scare me with their empty spaces*
> *Between stars—on stars where no human race is.*
> *I have it in me so much nearer home*
> *To scare myself with my own desert places.*

The really important change in our view of the Earth in space took place long ago when Copernicus removed the Earth from the centre of the universe. Since then modern cosmology has continued to put us in our place. We have now lost our central position, not only in space, but also in time. Since Darwin's day we have discovered that human beings, like everything else, are the products of a natural process, evolution, which stretches back for millions of years. On this enlarged time-scale the lifetime of the human race is only a minor part of the history of the universe. And that is not all. Modern cosmology tells us that we are not the only pebble on the beach; we must accept that probably, almost certainly, there are millions of other worlds.

These great changes in perspective have influenced our ideas about the significance of human life and our attitude towards established religion. For instance, the idea that there are other inhabited worlds may eventually prove to have been as influential as the recognition that the Earth is not at the centre of the solar system. For Giordano Bruno the plurality of other worlds was an inspiration which illustrated the infinite power of God, but in our own time it has made it harder for less devout people to accept the claims of historical religions, such as orthodox Christianity, which are based on unique events which took place on Earth. Needless to say, modern cosmology has not disproved these religions but it has altered the climate of belief; against our current picture of the universe much of what they tell us looks parochial. The old anthropomorphic ideas of religion look even less plausible than they did at the time of Newton's *System of the World*. In the words of Rupert Brooke, we can't help wondering if:

> *even fish*
> *trust—there swimmeth* ONE
> *who swam ere rivers were begun—*
> *squamous, omnipotent and kind.*

The modern vision of the cosmos, with its countless galaxies and stars, has encouraged the belief that the universe is indifferent to our fate. Despite the efforts of modern theologians to reformulate our ideas about God, it has redirected much of our religious interest away from questions of doctrine, such as the nature of the Trinity, to the more practical questions of good conduct. Science encourages us to ask only those questions to which we are likely to get an answer, and, preferably, an answer which can be tested by experience. If cosmology has removed the old Heaven then we must busy ourselves making a new one on Earth, at least until someone comes up with a better idea.

On a more constructive note, modern cosmology proclaims the unity of Nature, one of the main themes of modern science. Biologists tell us that all living things are made of the same sort of atoms as the 'dead' world around us, and now cosmologists tell us that man and his environment evolved from the same primordial cloud. Physicists go much deeper; they tell us that the basic properties of matter are not only universal, but can be understood only in terms of their interaction with the rest of the world and, in particular, with the observer who is making the observations. In short, science is telling us that everything is interrelated or, to borrow a phrase from the mystics, all Nature is one. This attitude to Nature has inspired many religions, particularly in the East, and is now gathering strength in the West. Our modern prophets of doom are always telling us that we must embrace the idea that all living things and their environment are part of the same eco-system, and that to survive such perils as over-population and nuclear war, we must learn to plan for all mankind as part of planet Earth. Against the background of modern cosmology their message looks realistic; perhaps this is the start of the better idea for which many of us are waiting.

As well as religion and philosophy, almost every other aspect of our culture has been influenced by our current vision of the universe. Take, for instance, the arts. The anthropomorphic cosmology of mediaeval times was so concrete that it lent itself to rich and powerful expression in painting, sculpture and drama. One cannot look at a mediaeval religious painting or watch a mystery play without realizing how *real* their ideas about Heaven and Hell were, and how remarkably effective they were as symbols of belief about the origin and purpose of the world. In contrast our modern picture of the universe is remote, abstract, and decidedly non-human. These qualities are certainly to be found in contemporary painting, music, and sculpture, and in many other fields as well. To discuss them in reasonable depth would be well beyond the scope of this chapter.

There is, however, one art-form in which the direct influence of cosmology can be seen at once—the vigorous literary genre of science fiction. The authors of science fiction have given to cosmology what it needs to compete with the old images of Heaven and Hell, not just human interest but superhuman (and subhuman) interest. Maybe it is a trick of the imagination, but I seem to recognize some of the demons from mediaeval pictures of Hell on the dust jackets of sci-fi books.

Why bother about science?

At the beginning of this chapter I asserted that it is not enough to value science only for its practical uses or as an ornament of society; it is a pillar on which our culture rests. Our attitude to this assertion must obviously depend upon whether or not we believe in the possibility of progress and on how we conceive of a better world. If we define a better world largely in material terms or take the view that any other form of progress is illusory, then it is reasonable to see the value of science as only twofold; as an aid to material progress and also, perhaps, as an intellectual pastime. This is a view which had its heyday in the 18th and 19th centuries. On the other hand if we believe, or hope, that society does progress on the ideal as well as on the material plane, then there is another important reason why we should value science.

Anyone who has read the trial of Galileo knows that human institutions tend to preserve ideas as a rock preserves fossils. If the ideas of a society are not to fossilize, but to progress, they must be both flexible and realistic, and to ensure this, society must welcome and encourage the pursuit of science. Scientific research is our major paradigm of progress; it continuously enlarges our knowledge of the world and of ourselves and, by so doing, it exerts a dynamic influence on culture which prevents ideas from going stale. Scientists are like a group of people doing a jigsaw puzzle. The pieces are facts and the picture is of the world as we find it to be, not as we might imagine or prefer it to be. When a new fact fits into the puzzle, and the picture still makes sense after it has been put in place, then that fact is accepted as being true. If our ideas about the world and ourselves are to remain realistic, then we must keep an eye on this picture as it grows; if they are to remain flexible we must be prepared to revise them in the light of the changing scene—a static vision of the world belongs to the Middle Ages.

But not everyone is willing to watch this picture. Most of us have read Aldous Huxley's *Brave New World* and seen the photographs of Hiroshima; we have heard about over-population, pollution, nuclear waste, energy shortage, genetic engineering, and so on, ad nauseam. Some of us, the more romantic, are inclined to turn our backs on science and hope that it will go away. But most of us know that our only hope of solving these awful problems and of making progress towards a better world depends, not simply on expertise but on vision and wisdom. To reach a better world there is no simple formula, we must always be making value-judgments in which we weigh profit, loss, freedom, justice, beauty, and truth against each other. If we are to succeed we shall need all the science we can get, not simply to solve our technical problems or to entertain us, but to guide us to ideas, values and actions, which are true—true in the sense that they are realistic and based on the best information available at the time about man and the world. Wisdom and vision both need imagination but it must be imagination informed by truth, the sort of truth which science offers. Most systems of ethics and morality place a high value on truth, but what they fail to recognize is that so few truths are eternal; our knowledge of what is true about man and the world changes as it grows. In Newton's day the existence of absolute space and absolute time were regarded as 'eternal truths'; not long ago Einstein showed that they were not absolute but relative.

There is something else about the search for wisdom which must be recognized—the importance of what a physicist might call *complementarity* in human understanding. It seems that there are many things, perhaps everything, in the world, which cannot be understood satisfactorily in terms of 'single vision'. For instance we can know about the piano either as a mechanical contrivance of hammers and strings or as the medium of a Beethoven sonata; but to understand the piano thoroughly—and, by the way, to improve it—we must know it in both ways at once. The same is true for the idea of the unity of Nature. We can approach it subjectively through religion or the arts, or we can approach it from the opposite direction, objectively through science. For a satisfactory understanding these two approaches must validate each other and, if they do, that is the beginning of wisdom. To object to science because it is 'reductionist' is, more often than not, absurd; 'reductionism' is a fancy word for finding out how things work, and to know how they work is usually an essential part of understanding them satisfactorily. To object to religion or the arts on the ground that they are 'subjective' is equally absurd; 'subjectivism' is a fancy word for seeing things in relation to ourselves. It is also an essential part of understanding them; as Wordsworth says, it is the poet's way of welcoming 'what is now called science . . . as a dear and genuine inmate of the household of man.'.

Seen in this light the pursuit of science is not, as I have said, simply an aid to material progress or an ornament of culture, it is one of the main pillars on which our hopes of a better world rest. At the risk of being misunderstood, we might call the pursuit of science one of our moral duties.

A moral duty can be very dull and, like any other occupation, a lot of science is dull and so, of course, are a lot of scientists. Humanists, who are apt to think that they have a monopoly of imagination, often present science as the enemy of poetry and imagination, hell-bent on removing mystery from the world. This attitude is, I suppose partly a hangover from the 19th century when science was particularly 'mechanical' and partly a confusion between the technology—computers and all that—and science of our own time. In actual fact the accusation that science is an enemy of poetry and imagination is, nowadays, wide of the mark. The real world which science explores has turned out to be strange, infinitely complex, and wonderful. The laws of physics are far more abstract, the mechanisms of heredity and immunity are far more intricate, snowflakes are far more beautiful than any writer, poet, or artist could ever have imagined. Indeed the scientific exploration of Nature nourishes our imaginations and is now one of our principal sources of new and beautiful ideas; it is a great adventure of the human spirit. If you are looking for enemies of the imagination, don't look at science, look at some of those topics among the humanities which have been cut off from science for far too long.

Again, the idea that science removes mystery from the world, like a vacuum cleaner removing cobwebs, is equally misleading. It is true that science removes minor mysteries, but in doing so, it shows us where the major mysteries really are. The art of scientific research is to ask questions about Nature which we can hope to answer, and to represent any remaining mysteries by symbols which work sufficiently well for the purpose in hand. Let me illustrate this by one example taken from physics. I choose physics because, in many fields, it has already ploughed through the early 'mechanical' stages of scientific enquiry to uncover mysteries which demand radically new ways of thinking about quite basic ideas, like cause and effect, or individual identity. Most topics in biology are, by comparison, in an earlier stage of development; no doubt they too will reach their own mysteries in due course, for instance, the problem of consciousness. In the meantime, let us look at what physicists have to say about the nature of light. Laboratory experiments show that light behaves sometimes as a wave, sometimes as a particle, depending on what sort of experiment we do. To cope with this rather baffling experience physicists have developed two sets of equations—one for waves and one for particles—which tell us what light will do in any particular case. These equations answer all our practical problems, but they don't answer the question—what is light *really like*? For years people have tried to answer this question, but nowadays they have given it up, because they suspect that it has no meaning. If we insist on an answer, then it is that 'light is like light'. If light is neither a particle nor a wave, then we must accept that it is a 'wave-particle', something which we cannot understand in terms of our everyday experience of things which we can touch and see. Physicists have shown us that light can only be understood in terms of what it does, and not in terms of what it *is*, which is one way of saying that it is a mystery.

If I may digress for a moment, something very similar has happened in other fields. Look, for example, at the much older problem of making a working model of the Trinity. The solution is given in the Athanasian Creed. The Trinity is described as three distinct Persons, each with His own function, co-equal, co-eternal, and each of them is God. But there is only one God, and so we must think of Him as being 'three Persons without dividing the substance'. In other words we are given a practical system of symbols for the Trinity, but its true nature remains a mystery. No doubt there are many differences between the triple nature of the Trinity and the dual nature of light; nevertheless, I find the similarity between the ways in which people have coped with these two mysteries instructive.

As for the great mysteries which stand in the shadows of all human thought, such as the origin and purpose of the world, modern science cannot be accused of sweeping them away. The mystery of Creation is intact, pushed back by 20 billion years perhaps, but, nevertheless, where it always was—in the beginning. Nor has science anything to say about the purpose of the world. It has told us a good deal about the scenario of the play but has left us to guess the plot. But in doing this it has changed our picture of the universe so much that it has made obsolete many of our religious symbols, especially those which represented the mystery of purpose. For many people this has created a spiritual wasteland, which is, unfortunately what happens when we hang on to ideas which are no longer realistic, or fail to reinterpret them in the light of what we now know.

How can we help science to play the important part which it should in our society? One obvious way is to promote a better understanding of what science is, how it is done, and by what sort of people. It goes without saying that society should be aware, as well as wary, of the results of scientific research or, at the very least, of what the research is about. The farther the average person is from understanding what modern science has to say, the more room there is in our culture for the 'cults of unreason', for everything from tarot cards to scientology. If science is no more intelligible than magic, people will prefer magic because it is usually much better presented. If too much of science is seen as being esoteric or irrelevant it will lose its influence on society, just as theology did before it. Indeed there are too many parallels for comfort between the organization and arcana of science and those of the Church, particularly of the earlier Church.

One of the main difficulties in popularizing science is, of course, translation. So much of science, for instance the Theory of General Relativity, is written in the language of mathematics and will inevitably remain largely unintelligible to most people; it is unfortunately not always possible to translate mathematical ideas into everyday language without a serious loss of meaning. Sad to say, the time is long past when we could learn our cosmology from a poet, such as Dante. In his day scientific works were written in Latin or Greek, and although few people could read then, they could be translated accurately. Nevertheless, we must do our best and encourage, not only the writing of popular science by people who really understand what they are writing about, but also the idea that some

The Earth seen from space. A photograph taken from the Apollo 17 spacecraft in December 1972. The ▷ view extends from the Mediterranean to Antarctica and shows most of the coast of Africa.

knowledge of mathematics and science is part of a 'cultured' person's stock-in-trade just as Latin and Greek used to be.

A more practical objective, one which we have a much better chance of reaching, is to promote the understanding of the *relevance* of science. Experience suggests that this can be done without having to explain the actual results of scientific research in boring detail. The relevance of science to the practical needs of society can be easily explained; the relevance of one science to another and the relevance of science as a whole to our culture are both more difficult to convey, but we must try. We must make it clear that the scientific vision of the world is neither a rival nor an alternative to any other point of view, but is an essential part of learning to be at home in this mysterious universe.

Finally, whatever we do, let us not forget what Walt Whitman said:

When I heard the learn'd astronomer,
When the proofs, the figures, were ranged in columns before me.
When I was shown the charts and diagrams, to add, divide, and measure them,
When I sitting heard the astronomer where he lectured with much applause in the
 lecture-room,
How soon unaccountable I became tired and sick,
Till rising and gliding out I wander'd off by myself,
In the mystical moist night air, and from time to time,
Look'd up in perfect silence at the stars.

Glossary

altitude The angle of elevation of an object above the horizon as measured along a great circle through the object perpendicular to the horizon.

amino acids Protein molecules of living matter are made from twenty different complex molecules called amino acids.

astrolabe A obsolete type of instrument for measuring the angle of elevation (altitude) of objects such as the sun, stars, etc. It consists of a circular dial engraved with a scale of degrees and an index arm (alidade) pivoted in the centre of the disc. This arm is lined up with the object by some simple sighting device and the altitude of the object is then read off the circular disc.

atomic clock A clock whose rate is controlled by the frequency of radiation from the caesium atom.

atomic second/(S.I. second) A second based on the frequency of the radiation from a caesium atom. The Système Internationale (S.I.) second is defined as 9 192 631 770 periods of the caesium atom.

atomic time A system of time based on counting the number of vibrations (transitions) of the caesium atom.

black hole A star which has collapsed under the force of gravity to such a high density that no radiation can escape from it.

caesium A soft metal. An element with atomic number 55.

celestial equator The projection of the Earth's equator on to the sky. A great circle on the celestial sphere 90° from the celestial poles.

celestial sphere An imaginary sphere surrounding the Earth on which the position of objects in the sky can be mapped.

circadian rhythm A cycle which is repeated at daily intervals— approximately every 24 hours.

declination One of two measurements— the other is right ascension—which fix the position of an object on the map of the sky. It is the angular distance of the object from the celestial equator, measured in degrees along a line which passes through the object and is perpendicular to the celestial equator.

Doppler effect When a source of radiation (sound, light, or radiowaves) is moving towards or away from an observer the apparent frequency of the radiation is changed due to their relative velocity. It is increased when they are approaching, and decreased when they are separating.

ecliptic The plane of the Earth's orbit around the sun projected on to the sky—the apparent annual path of the sun on the celestial sphere.

electromagnetism The topic of electromagnetic fields and waves. An electromagnetic wave (light, radio waves, X-rays, etc.) is a disturbance in a system of crossed electric and magnetic fields which propagates with the speed of light.

ephemeris Table showing the positions of a celestial body at various times

ephemeris time A system of time based, not on the rotation of the Earth, but on the orbital motions of the solar system.

equinox The time (or the position in the sky) when the annual path of the sun along the ecliptic crosses the celestial equator. This happens twice a year, at the vernal equinox (about 21 March) and at the autumnal equinox (about 23 September) and at those times the day and night are of equal length.

escapement The mechanism in a clock which enables the pendulum or hairspring to control the rate of the clock.

galaxy A star system containing as many as 100 billion (10^{11}) stars.

gnomon The rod or pillar of a sundial whose shadow shows the hour.

gymbals A system of supporting chronometers, magnetic compasses and other instruments which remains horizontal on a moving ship: the system employs two freely mounted axes at right angles to each other.

heliocentric Centred on the sun.

ionized An atom which has gained or lost an electron is called an *ion*. A gas in which there are many such atoms is an *ionized* gas.

light-year The distance travelled by light in one year: 9·46 million million kilometres.

lunisolar year A year based on both the sun and moon.

magnetosphere A region surrounding the Earth in which charged particles are trapped in the Earth's magnetic field. It extends roughly from 1000 km to 60000 km from the Earth's surface.

mantle of the Earth The solid, rocky part of the Earth immediately below the crust.

mean solar day The interval between successive passages of the mean sun across the observer's meridian; the average length of a day measured by observing the real sun for one complete year.

mean solar time There are 24 mean solar hours in one mean solar day. These hours are divided into mean solar minutes and mean solar seconds.

mean sun An imaginary sun which moves at a perfectly constant speed along the celestial equator taking one year to make a complete orbit.

meridian A great circle on the sky (celestial sphere) which passes through the zenith and both the celestial poles. Also a great circle on the Earth which passes through the observer and both the poles.

Metonic cycle A cycle of 19 years after which the moon's phases recur on the same days of the solar year.

micrometer eyepiece An eyepiece for a telescope which contains two parallel wires whose separation can be adjusted by means of a screw-thread and dial. By aligning the wires with two different stars, the angular separation of the stars can be measured very precisely.

microsecond An interval of time; one millionth (10^{-6}) of a second.

minute of arc A measure of angle. 1/60 part of a degree.

momentum The product of the mass and velocity of a body.

nanosecond An interval of time: one thousand millionth (10^{-9}) of a second.

neutron A subatomic particle with no charge and a mass approximately the same as a proton.

nova A star which erupts and throws off a small fraction of its mass into space. It becomes suddenly brighter by many hundreds or thousands of times and then fades back to its original state.

parallax The apparent change in the direction of an object when viewed from different positions.

photon The smallest possible unit of radiation (e.g. light). A quantum of radiation.

178

precession A slow gyration of the Earth's axis which causes the position of celestial poles to sweep out a circle with a radius of $23\frac{1}{2}°$ in 26 000 years.

precession of the equinoxes A slow westward movement of the position of the equinoxes along the ecliptic due to the precession of the Earth's axis.

proton A sub-atomic particle carrying unit positive charge. A component of the nucleus of an atom.

pulsar An object which is observed to emit short pulses of radio waves and also, in two or three cases, pulses of light. They are believed to be rotating neutron stars.

quadrant An instrument with an arc of 90° used for measuring the angle of elevation (altitude) of the Sun, stars etc.

quartz oscillator An electrical oscillator whose frequency is controlled by the natural frequency of vibration of a block of quartz. The quartz is made to vibrate by an alternating electric field.

quasar This term is short for quasi-stellar radio source: quasars are extremely powerful radiators of light and radio waves and can be observed at very great distances. The origin of their radiation is not fully understood.

radar A system of locating objects by transmitting radio waves and observing the reflected waves.

red-shift The displacement in the lines (colours) of the light in the spectrum of a distant star or galaxy towards longer wavelengths: usually attributed to motion of the star or galaxy away from the observer.

right ascension One of the two measurements—the other is declination—which describe the position of an object on a map of the sky. It is measured along the celestial equator in hours, minutes and seconds from the First Point of Aries where the path of the sun crosses the celestial equator at the vernal equinox.

rubidium A soft metal. An element with atomic number 37.

saros A cycle of similar eclipses that recurs at intervals of 6585 days or roughly 18 years.

second of arc A measure of angle. 1/3600 part of one degree.

sidereal day/noon The interval between two successive passages of the same star across the meridian at any place: the true period of the Earth's rotation: about 23^h 56^m 4^s of mean solar time. The sidereal day starts when the First Point of Aries (the point which marks the vernal equinox) crosses the meridian. This moment is 0^h 0^m 0^s sidereal time and is called sidereal noon.

sidereal time Time measured relative to the stars and not the sun. Each sidereal day is divided into 24 sidereal hours (minutes and seconds) which are therefore shorter than solar seconds.

solstice Positions on the celestial sphere (or times) when the path of the sun (the ecliptic) is at its greatest distance from the celestial equator. The summer solstice corresponds to mid-summer and the longest day; the winter solstice corresponds to mid-winter and the shortest day.

spectrum The pattern of colours formed when light is dispersed (e.g. by a prism) into its component colours.

spectroscope An instrument for dispersing or splitting up light into its component colours by means, for example, of a prism.

spherical triangle A triangle drawn on the surface of a sphere.

supernova An exploding star which throws off its mass into interstellar space. For a short time it is intensely bright.

synodic month The time which it takes for the moon to complete a cycle of phases (e.g. the time from one new moon to the next).

Glossary

temporal hours Hours defined by dividing the intervals between sunrise and sunset into twelve hours of daylight and twelve hours of night.

tropical year The time taken by the Earth to make one complete orbit around the sun, measured as the interval between two successive vernal equinoxes.

Universal Coordinated Time (U.T.C.). The basic system of time in general use throughout the world at present. The basic unit is the standard second (S.I.) derived from an atomic clock. The system is kept in step with mean solar time (U.T.) by occasional 'leap seconds'.

Universal Time (U.T.) Mean solar time at longitude 0° i.e. at Greenwich. Formerly known as Greenwich Mean Time (G.M.T.).

verge and foliot escapement An early form of clock escapement, consisting of a horizontal rod carrying two weights (the foliot) and carried by a shaft (the verge) with two pallets which engage with a toothed wheel.

white dwarf An old star that has exhausted its supply of nuclear energy and has collapsed into a small extremely dense body.

zenith The point on the celestial sphere directly above the observer.

180

Further reading

* indicates books for the general reader

General astronomy

Exploration of the universe, by G. Abell (Holt, Rinehart and Winston, New York, 1964).

Essentials of astronomy, by L. Motz and A. Duveen (Blackie, Glasgow, 1966).

New horizons in astronomy, by J. C. Brandt and S. P. Maran (W. H. Freeman, San Francisco, 1972).

History

A history of astronomy from Thales to Kepler, by J. L. E. Dreyer (Dover Publications, New York, 1953).

A short history of scientific ideas to 1900, by C. J. Singer (Oxford University Press, 1959).

The origins of modern science (1300–1800), by H. Butterfield (G. Bell, London, 1949).

The fabric of the heavens, by S. Toulmin and J. Goodfield (Hutchinson, London, 1961).

Augustine to Galileo (science in the late Middle Ages and early modern times) 2 vols. by A. C. Crombie (Heinemann, London, 1961).

The mechanization of the world picture, by E. J. Dijksterhuis (Oxford University Press, 1961).

The place of astronomy in the ancient world, edited by F. R. Hodson (Royal Society of London, 1974).

The exact sciences in antiquity, by O. Neugebauer, (Brown University Press, Providence, R.I., 1969).

A history of ancient mathematical astronomy (3 vols), by O. Neugebauer (Springer-Verlag, New York, 1975).

History of the theories of aether and electricity, 2 volumes, by E. Whittaker (Thomas Nelson, Sunbury-on-Thames, 1953).

William Herschel, M. A. Hoskin (Oldbourne Press, London, 1963).

Greenwich Observatory, by E. G. Forbes (Taylor and Francis, London, 1975).

The making of the modern mind, by J. H. Randall (Houghton Mifflin, London, 1940).

The discarded image, by C. S. Lewis (Cambridge University Press, 1964).

The alchemists, by F. Sherwood Taylor (Paladin, St. Albans, 1976).

Time and the calendar

Chronology of the ancient world, by E. Bickerman (Thames & Hudson, London, 1968).

Explanatory supplement to Astronomical Ephemeris (Her Majesty's Stationery Office, London, 1961).

Time and the calendars, by W. M. O'Neil (Sydney University Press, 1975).

The marine chronometer, by R. T. Gould Holland Press, London, 1923

Sundials, by F. W. Cousins (John Baker, London, 1969).

Stonehenge decoded, by G. S. Hawkins (Dell, New York, 1966).

The week, by H. F. Colson (Cambridge, University Press, 1926).

Navigation

The art of navigation in Elizabethan and early Stuart times, by D. W. Waters (Hollis & Carter, London, 1958).

A history of marine navigation, by Per Collinder (Batsford, London, 1954).

The haven-finding art, by E. G. R. Taylor (Hollis & Carter, London, 1956).

The birth of navigational science, by E. G. Forbes (Maritime Monograph No. 10, National Maritime Museum, Greenwich, London, 1974).

Cook the navigator, by J. C. Beaglehole (*Proceedings of the Royal Society*, A **314**, 27, 1970).

Captain Cook and the transit of Venus of 1769, by Sir Richard Woolley (*Notes and Records of the Royal Society of London*, **24**, No. 1, 1969).

We the navigators, D. Lewis (Australian National University, 1972).

Cosmology and relativity

The measure of the universe: a history of modern cosmology, by J. D. North (Oxford University Press, 1965).

Principles of cosmology and gravitation, by Michael Berry (Cambridge University Press, 1976).

Relativity, by A. Einstein, authorised translation by R. W. Lawson (Methuen, London, 1960).

Index

Index

Subjects